W9-BSS-086

Conquering Infertility

ALSO BY Alice D. Domar, Ph.D.

Self-Nurture (with Henry Dreher)

Healing Mind, Healthy Woman (with Henry Dreher)

Six Steps to Increased Fertility
(with Robert Barbieri and Kevin Loughlin)

Conquering Infertility

DR. ALICE DOMAR'S GUIDE TO ENHANCING FERTILITY AND COPING WITH INFERTILITY

Alice D. Domar, Ph.D.,

AND Alice Lesch Kelly

VIKING

VIKING
Published by the Penguin Group
Penguin Putnam Inc., 375 Hudson Street,
New York, New York 10014, U.S.A.
Penguin Books Ltd, 80 Strand, London WC2R 0RL, England
Penguin Books Australia Ltd, 250 Camberwell Road, Camberwell, Victoria 3124, Australia
Penguin Books Canada Ltd, 10 Alcorn Avenue,
Toronto, Ontario, Canada M4V 3B2
Penguin Books India (P) Ltd, 11 Community Centre, Panchsheel Park, New Delhi—110 017, India
Penguin Books (N.Z.) Ltd, Cnr Rosedale and Airborne Roads, Albany, Auckland, New Zealand
Penguin Books (South Africa) (Pty) Ltd, 24 Sturdee Avenue,
Rosebank, Johannesburg 2196, South Africa

Penguin Books Ltd, Registered Offices:
Harmondsworth, Middlesex, England

First published in 2002 by Viking Penguin,
a member of Penguin Putnam Inc.

10 9 8 7 6 5 4 3 2 1

Library of Congress Cataloging-in-Publication Data

Domar, Alice D.
Conquering infertility : Dr. Alice Domar's guide to enhancing fertility and coping with infertility / by Alice D. Domar and Alice Lesch Kelly.
p. cm.
ISBN 0-670-03155-0
1. Infertility. 2. Fertility, Human. 3. Human reproductive technology. I. Kelly, Alice Lesch. II. Title.

RC889 .D597 2002
616.6'9206—dc21

2002025884

This book is printed on acid-free paper. 

Printed in the United States of America
Set in Goudy
Designed by BTDnyc

To Katie
—A.D.D.

To my mother, Catherine Walsh Lesch,
and in memory of my father, George A. Lesch Jr.
—A.L.K.

Authors' Note

Although this book is a collaborative effort by Alice D. Domar, Ph.D., and Alice Lesch Kelly, for the sake of clarity and simplicity we use the pronoun "I" for Dr. Domar (Ali) throughout. The scientific findings and case histories in this book are a product of her research and clinical work. The names and identifying details of the patients discussed in this book have been changed to protect their privacy.

The advice in this book is offered as a health recommendation, not as a substitute for standard medical care. We recommend that you use mind/body approaches only in conjunction with—not instead of—mainstream infertility care. Consult your physician about any symptoms or illnesses you may have, and inform him or her about any treatments upon which you embark.

For seventeen years I was at the Mind/Body Medical Institute at Beth Israel Deaconess Medical Center, which is where I did all of my research and started the Mind/Body Programs for Infertility. As we were completing this book, however, I accepted an offer to become the director of the Mind/Body Center for Women's Health at Boston IVF. The infertility programs are moving to Boston IVF, where they will continue and be expanded.

Acknowledgments

Thank you to all of the people who have contributed their time and energy to the Mind/Body Infertility Program and to this book. These include Alan Penzias, M.D.; Kim Thornton, M.D.; Barbara Nielsen, M.Div., Ph.D.; Ellen Sarasohn Glazer, LICSW; Diane Clapp, R.N., as well as the people on my staff who have been infertility group leaders: Patty Martin Arcari, Ph.D., R.N.; Eileen O'Connell, Ph.D.; Janet Fronk, R.N., M.S.; Leslee Kagan, R.N., M.S., N.P.; Harriet Redmond, R.N., M.S., C.S.; Ellen Slawsby, Ph.D.; Melissa Freizinger, M.A.; and Angel Seibring, Ph.D. I would also like to thank Christine Driscoll, Michele Chausse, Carol Tennant, and Eva Selhub, M.D.

I would also like to acknowledge the brave journeys that all of my patients have made and how much they have taught me over the years.

Many thanks to my patients—you know who you are—who read early drafts of this book and gave such wonderful, constructive feedback. I thank my colleagues in the mental health professional group of the American Society of Reproductive Medicine for their support and teaching over the years. My appreciation goes also to all of the infertility doctors and nurses in the Boston area who have referred patients to me over the years. My deepest gratitude goes to Joseph Mortola, M.D., my co-investigator on the federal infertility study, and to Bruce Kessel, M.D., who took over after Joe died. I would like to thank Herbert Benson, M.D., and Marilyn Wilcher for their consistent support of the infertility program while they were at the Mind/Body Institute. And my most sincere appreciation to Alan Penzias, M.D., for his encouragement in moving the programs to their new home at Boston IVF, and to Michael Alper, M.D., Ronald P. Jones, Benjamin Sachs, M.D., the Boston IVF board of directors, and the rest of the Boston IVF staff for making the move a reality.

Tremendous thanks also to my husband, Dave, and my daughters, Sarah and Katie, for their patience while I worked on this book.

—*Alice D. Domar, Ph.D.*

I would like to express my deepest gratitude to the Lesch and Kelly families, who have always been so supportive of my writing. Thanks also to the friends and fellow writers in my e-mail village—I can't imagine surviving the long days at my desk without messages of encouragement and friendship and silliness from Tim Gower, Carol Hildebrand, Kim Nash, Christopher Lindquist, Victoria Abbott Riccardi, Sarah Bowen Shea, and the members of my Boston-area freelance writers' group. A special thank-you to Nancy Gottesman, who assigned me my first freelance story and introduced me to Ali Domar. An enormous hug to my sons, Steven and Scott. And most of all to my husband, Dave, who insists that it's neat being married to a writer even though it occasionally makes for a rather chaotic life. I love you all.

Warmest thanks also to the many infertility patients who shared their stories with me. During the months that I conducted most of my interviews with Ali's patients, my father was fighting a seventeen-month battle with cancer. The women and men I spoke to showed such strength of character, resilience, dignity, and goodness that simply listening to their stories helped me feel stronger and better able to cope with my father's illness and death. What an amazing group of people. It was an honor to interview you.

And thank you, Ali, for asking me to be your coauthor. It has been a tremendous experience for me and a pleasure working with you.

—Alice Lesch Kelly

We would both like to thank the dozens of Ali's patients who shared their stories for this book. We know that it can be difficult to disclose personal stories, but these women and men did so willingly in order to help other people who are struggling with infertility. Thank you for your generosity and kindness.

We are also indebted to our agent, Christine Tomasino, who waited ten years for this book and who has been an enthusiastic, supportive friend every step of the way; and to our editors, Pam Dorman and Susan Hans O'Connor, for their diligent, compassionate, and thorough critique of the manuscript, as well as to the rest of the Viking Penguin team.

Contents

Introduction

About ten years ago, at a time in my career when I was completely entrenched in infertility both in terms of the patients I was treating and the research I was conducting, a good friend called to tell me that she was going to have a baby. Without thinking, I replied, "That's wonderful! So how did you get pregnant?" My mind was on Clomid, IUI, and IVF. "What do you mean, how did I get pregnant? The same way *everybody* gets pregnant!" It turns out she'd conceived the old-fashioned way and was shocked thinking that I was asking about the intimate details of what position they'd used, which room of the house they'd conceived in, and so on. It hadn't occurred to my friend that someone would get pregnant any other way than by having unprotected sex. And it hadn't occurred to me, at first, that someone could actually get pregnant simply by having sex.

Does this sound familiar to you?

If it does, and if you are thinking these days that sex and pregnancy seem to be completely unrelated, you need to read this book.

If you have to get off the phone quickly so you can hide your tears when your best friend calls to say she's pregnant, you need to read this book.

If you find yourself constantly becoming angry at your mother for not "getting it," or you worry about your reaction the next time someone tells you to just relax and you'll get pregnant, or you've lost interest in your job, or you can't stand it that everyone you know is pregnant, you need to read this book.

If you feel sad more than you feel happy, if you find yourself wondering if the guy you married has a sensitive bone in his body because *his* life seems to go on even after you get your period, if you find you worry constantly about seeing a pregnant woman on the street or being surprised by a pregnancy announcement, if you really wonder whether you will ever be happy again, you need to read this book.

Infertility stinks. It can be an all-encompassing heartbreak that leaves you feeling isolated, depressed, angry, and hostile. But I'm here to tell you that you don't need to feel that way. You can learn to pursue infertility treatment without feeling as if your life and body have been taken over. You can learn to talk with your husband calmly and rationally and to make decisions that are best for both of you. You can learn how to decrease the stress-related symptoms such as insomnia and headaches that further decrease the quality of your life. And believe it or not, you can actually learn to laugh during infertility.

This book is designed to give you back your life. As you read along, I want you to think, "I thought I was the only one who felt like that!" or "I thought my husband and I were the only ones reacting to infertility in such an unhealthy way!" or "I feel as if you know me—my situation is being described on every page!" And I want you to come to the realization that you are going to be OK, that there are *lots* of things you can do to help yourself feel better. Infertility can seem like the ultimate loss of control. In this book I will teach you how to take control of your infertility, your body, your relationships, your mind, and your soul.

This book is the result of sixteen years of work on infertility. I began doing research on the impact of stress-reduction techniques on infertility in 1986, and started my first mind/body infertility group in 1987. I published my first paper summarizing my research in 1990. I published two books on stress reduction for women and cowrote, along with two physicians, a book on infertility in which one chapter is dedicated to the mind/body connection. But it really wasn't until now that I felt ready to put all those years of experience conducting research and group and individual therapy into book form. Before I wrote a book on this subject, I wanted to have conducted the research and thoroughly examined all the literature on the connection between stress and infertility. I wanted to treat as many patients as possible so that I could reference stories that would have meaning to every reader. I wanted to be able to say that I think I've heard everything—all the myths surrounding infertility, the insensitive comments people make, the desperate actions patients take, and the ins and outs of modern medical treatment.

Finally, I wanted to learn enough from my research and clinical practice to give infertility patients hope. And now I can do that. This I know for sure, based on sixteen years of research and clinical work with thousands of infertility patients:

You will be happy again. Life will become joyful again. And somehow, some way, if you want to become a parent, you will.

People often ask me why I've made infertility my life's work. It all started with my mother and the tremendous impact her infertility had on my family. In the 1950s, when my mother had trouble conceiving, her doctors were unable to tell her why she wasn't getting pregnant, and they had little to offer her in the way of treatments. These days medical science has so many more answers for infertile women. Reproductive endocrinologists can count sperm, measure its motility, determine the time of ovulation, evaluate the health of eggs, analyze hormone levels, and inspect the fallopian tubes, ovaries, and cervix for abnormalities. Once a suspected problem is uncovered, doctors can often correct it. Surgery can remove cysts and obstructive scar tissue, and drugs can boost hormone secretion and spur ovulation. If conception still fails to occur, physicians can offer infertile couples a range of assisted-reproductive technologies that the women in my mother's generation could never have dreamed of.

But with all of today's wonderland of technological treatments for infertility, something important is missing. Despite the many medical achievements of the past decades, very little progress has been made in treating the emotional side effects of infertility. When my own mother was trying to become pregnant, she felt depressed, anxious, and isolated. She was left to bear the emotional strains of infertility alone. Amazingly, that continues to be true of many infertile women today. In just about every major city in the world, you can find clinics full of doctors who can diagnose and treat infertility. But it's much harder to find a physician who will help you cope with the anger, jealousy, sadness, confusion, and profound yearning that accompany infertility.

After years of trying, my parents finally conceived my sister. Five and a half years later, after they had pretty much given up the dream of having a second child, I arrived. And although my parents certainly didn't dwell on their experience with infertility, I grew up with an acute awareness of the effect it had had on them and on us as a family. My parents frequently talked about how infertility had affected them—and about how much they treasured my sister and me because they had almost abandoned any hope of having children.

I believe that my mother's experience with infertility indirectly helped shape my career choices. As the director of the Mind/Body Center for Women's Health at Boston IVF, I design and lead mind/body workshops that have helped thousands of women cope with the stress of infertility. These programs, which have drawn participants and media attention from around the world, teach infertile women dozens of relaxation techniques, stress-management strategies, and other mind/body skills that help alleviate the strain of coping with infertility and boost pregnancy rates.

My journey into mind/body medicine started early. As a child, I always wanted to be a doctor, but as I got older, I also knew that I was particularly interested in helping people emotionally. When I finished college, I couldn't decide whether I could better reach my goal as a psychologist or as a physician. So I took two years off to make up my mind. During that time I worked for two doctors: a neurologist and a psychologist who were doing brain research at Children's Hospital in Boston. I figured that working for both a psychologist and a neurologist would help me decide which career path to take.

During those years it became very clear to me that although I was interested in medicine, what really fascinated me was the emotional aspect of disease. As I looked at brain scan after brain scan, I was drawn not to diagnosing pathology but to helping the people whose scans I was studying. I wanted to meet the patients and families who were struggling to cope with disease and help them survive its emotional punch. So I chose to become a psychologist, and I enrolled in the health psychology program at the Albert Einstein College of Medicine/Ferkauf School of Professional Psychology of Yeshiva University in the Bronx. When it came time to choose medical specialties, I picked OB/GYN. Amazingly, I was the first person in the history of my health psychology program to choose this specialty—which says a lot about how little attention had been paid to the connection between mind and body in the field of women's health.

My twin interests in women's health and health psychology led me to the Mind/Body Medical Institute in Boston, where I began working with Dr. Herbert Benson, the father of mind/body medicine. That's when everything came together for me, and I realized that I could use my medical expertise, my psychological training, my personal life lessons, and my empathy to help women to cope with both the psychological and physiological challenges presented by their medical conditions.

Over the years, as I have worked on mind/body clinical programs and research projects, I have recognized something that is crucial when it comes to successfully treating women's medical conditions: *Women's minds must be treated along with their bodies*. That is true no matter what causes their suffering, whether it is infertility, cancer, insomnia, chronic pain, AIDS, or menopause. To separate mind from body, to treat one without attending to the other, is foolish and ineffective. When you treat a woman's mind as well as her body, almost without exception she feels better and can cope more effectively with her condition. And in some cases her symptoms and prognosis improve.

In case after case after case we find that the women who learn to use mind/body strategies to manage the stress of infertility dramatically reduce their levels of depressive symptoms, anxiety, anger, and frustration. When infertile women learn to use simple techniques that allow them to relax, transform negative thought patterns, express their emotions, and develop strong sources of social support, they begin to take control of their lives again. They feel significantly less isolated and desperate. They sleep better, their relationships with their husbands improve, and they suffer from far fewer stress-related ailments such as gastrointestinal problems or insomnia. What's more, as I have shown in a number of studies published in peer-reviewed medical journals, they are more likely to get pregnant.

But getting pregnant is not the focus of this book. This book is about how infertile women can reestablish control over their lives. What feels most threatening to women experiencing infertility is being out of control. They can't control their fertility, they can't control their emotions, and in many cases they can't even control when they're going to make love to their husbands. They feel that their lives are falling apart, their careers are in shambles, and their personal relationships have withered—all because infertility has chipped away at their emotional health. The mind/body approach that I teach in my classes and in this book helps women regain control. Sometimes it helps them get pregnant, too—but making infertile women feel better is my primary goal. Pregnancy, if it occurs, is just a happy side effect.

In this book I'll provide you with an experience that is as close as possible to actual participation in one of my ten-session mind/body infertility programs. Central to my approach is the "relaxation response," our inborn capacity to reduce internal stress. I'll explain a variety of easy-to-learn ways to elicit the relaxation response, including meditation, mindfulness, yoga, body scan, progressive muscle relaxation, and autogenic training. I'll also show you the value of dozens of lifestyle changes, thinking strategies, and social approaches that you can use to help you deal with infertility. Not all of them will be right for you, and that's okay. When it comes to mind/body coping and relaxation strategies, I like to think that I'm offering a buffet of choices. The first time you go up to the table, you try a little of everything, and the next time you take just what you like. Here's an example: For some people meditation is a godsend—several of my patients say meditation has saved their lives. But others can't quiet their minds enough to enjoy meditation, and for them a yoga class is a far more effective way to bring peace to their agitated psyches.

I'll also take a close look at the connection between depression and infertility. Evidence is mounting that not only are depressive symptoms a very common side effect of infertility, but they may also impede your chances of getting pregnant as well. I'll help you determine whether you are depressed and, if you are, give you a variety of strategies to alleviate your symptoms.

In addition, I'll discuss the many issues that can cause emotional turbulence for infertile couples, including finances, difficulties with health-care providers, the differences between male and female approaches to facing infertility, the toll infertility can take on your career, lack of understanding among family and friends, jealousy, anger, and spiritual concerns.

You'll hear the voices of many women who have struggled with infertility. These women have endured—and survived—infertility by using mind/body techniques to help bring joy back into their lives. They have worked hard to strengthen their marriages, rebuild friendships, revitalize their careers, and make peace with their situations. Some have been wonderfully successful, and others continue to struggle. Most have become parents; others are learning to accept being childless. All of them are truly inspiring, and I hope that by reading their stories you'll not only learn the power of mind/body techniques but feel less isolated as well. Perhaps you'll even feel inspired to reach out to join a support group with other infertile women.

You'll also meet individuals who have decided to take less-traveled roads to parenthood. Some of them have chosen to adopt children, from either the United States or abroad, and others have become parents with the help of donated eggs or sperm. You may think that these options aren't right for you, but I urge you to put those thoughts aside and come back to them later. Many of the women who begin my program are hell-bent against anything but genetic, biological motherhood. But as they use mind/body techniques to restructure their thinking and to open lines of communication with their spouses, they sometimes discover that those other options are more palatable than they had originally thought.

Mind/body techniques aren't a magic wand that will make you pregnant. But they are an excellent, effective way to help you take back control of your life, cope with your infertility in a much more positive way, and prepare yourself to make choices that will contribute to your happiness and good health for the rest of your life.

CHAPTER ONE

Infertility, Stress, and Depression

When I logged on to my computer this morning, I took a quick look at my e-mail and found the usual collection of messages: a memo from a co-worker, a meeting reminder, a note from my sister, and some junk mail peddling stock tips that could make me rich—I wish! Then I saw a message from my good friend Cathy. The subject line, BAD NEWS, jumped out at me, so I opened the message and read it immediately.

"Got my period this morning. ☹!"

Her message was just five words—six if you include the doodad—and yet it told me so much. Even though Cathy didn't say "I feel so depressed!" or "What are we doing wrong?" or "Why is this happening to us?" I knew she was probably thinking these things. I knew that she most likely had cried her eyes out when those first few drops of blood of her period appeared, and that it probably took all the energy she could summon just to drag herself to work. And I know that if she sees a pregnant woman today, or hears a baby cry, or glimpses a picture of an infant on a co-worker's desk, her tears will return. When she gets home from work tonight, she's likely to snap at her husband, skip her workout, and spend the rest of the evening on the couch numbing herself with junk food and junk TV, trying to forget how bitterly disappointed she is that yet another month has gone by and she's still not pregnant.

I know this because I've seen it happen thousands of times.

Being unable to get pregnant is one of the most stressful things a woman can go through. Most of us, until we start trying—and failing—to get pregnant, assume that if and when we want children, we'll have them. As little girls we rock dolls in our arms and pretend to be mommies. As we grow up and become sexually active, we walk a shaky tightrope, assuming that the slightest slip could plunge us into an unwanted pregnancy. Yet we also feel completely confident that if we are smart about contraception we'll maintain complete control of when we will or won't get pregnant—we believe that it's all solidly in our own hands.

As newlyweds we think about when we'll start "trying," and we chat endlessly with girlfriends and sisters about whether it's better to give birth in spring or summer and which we'd rather have first, a girl or a boy. Then, once we finally do go off the Pill or toss aside the diaphragm or leave the condoms in the nightstand drawer and set out to make a baby, it's nothing but fun. A little champagne, some candles, some sexy lingerie, and after a few thrilling nights of unprotected lovemaking, we fully expect to be well on our way to a darling little baby. "After all, I don't shoot blanks," our husbands boast playfully. And as we wait for that first period not to arrive, we smile conspiratorially at women with babies and then march confidently off to the drugstore for a pregnancy test, happily anticipating a plus sign.

And then, for some women, nothing happens.

So you try again—but with the tiniest sliver of worry. You may pay more attention to the calendar and plan some extra midcycle sex. You nix the champagne and pop a few extra vitamins instead. But still, the next month, nothing happens. So you buy ovulation kits and cut out caffeine and ask friends for advice. You may exercise less (or more), eat less (or more), and insist that your husband wear boxers instead of briefs—and tough luck if they feel bunchy. "Deal with it," you think. You wonder whether you should make an appointment with your OB/GYN, or perhaps even a specialist. You fixate over what you could possibly be doing wrong. You have sex constantly. And yet your period keeps arriving, right on schedule.

Getting pregnant can start to become an obsession. As you fail to conceive, cycle after cycle after cycle, your anxieties may begin to haunt you, as negative thoughts loop endlessly through your mind. You blame yourself, your body, for failing, even though it may well be your husband's body that is the source of the problem. The content of those negative thoughts differs from woman to woman, but they're all related, a laundry list of should-haves and shouldn't-haves. *We should have started trying earlier. I shouldn't have drunk so much in college. My husband shouldn't have experimented with pot. I shouldn't have had an abortion in my twenties. I should have taken better care of myself.* Eventually your relationship with your husband starts to suffer. The thrill of frequent sex has worn off, and when your husband comes home from work exhausted on day twelve of your cycle, you tell him that you don't care *how* tired he is, he's doing it tonight if it kills him. You're panicked about not being able to conceive, but he's laid back. *Don't worry, he tells you. It will happen. Just relax and stop obsessing about it.* But you can't.

Then your best friend gets pregnant. She calls, all excited, prattling on and on about the names she's picked out and the darling crib she wants to buy and how excited her parents were to find out they're going to be grandparents. You pretend to be happy for her, but deep down inside you're insanely jealous, and you can't get off the phone fast enough. You're racked with guilt. You find yourself avoiding her and everyone else who has children. You just can't bear facing them.

You are stressed out. You may feel depressed, anxious, or angry. You might have trouble concentrating at work, and you may even cry every day. You begin to wonder if you'll ever have a baby, and if you'll ever be happy again. Your whole world is falling apart—just as it did for my patients Brenda and Janine.

Brenda's Story

Brenda, thirty-five, had been trying to have a baby for three years. She conceived naturally three times, but each pregnancy ended in miscarriage.

Then she couldn't even get pregnant for about a year, despite infertility-drug treatment and several intrauterine inseminations (IUI). "I was extremely depressed, although it's only in hindsight that I realize how depressed I was," Brenda says. "My husband kept telling me to see someone, to take something—but I never wanted to see anyone, because I was afraid they'd try to put me on an antidepressant, and since I was trying to get pregnant, I didn't want to do that. And I didn't want to see a therapist, because I knew what was wrong with me: If I could just have a baby, I'd be happy. I didn't need to go sit and whine in someone's office about not having a baby. I just knew that if I had a baby, I'd be happy, and if I didn't have a baby, I wouldn't be happy."

As time went on, Brenda became more depressed. "I wouldn't buy furniture or clothes, or I wouldn't plant flowers in the garden in the springtime—I wouldn't do anything I loved to do. I felt that until I had that baby, I couldn't do anything else. I was paralyzed and frozen. It was really hard for my husband to see me miserable all the time. He felt so helpless—he'd try to buy me things or do things for me, but nothing made me feel better."

Janine's Story

"I totally thought I'd get pregnant right away—my mother always got pregnant at the drop of a hat. Her nickname was 'Fertile Myrtle,'" says Janine, a now-forty-four-year-old adult-education instructor who started trying to conceive on her honeymoon. "I was so shocked that first month when I got my period." After six months Janine still was not pregnant. That's when it started to dawn on her that she might not be fertile. "I started reading, investigating, looking at my options, and talking to people."

During the following months she underwent a raft of painful tests, procedures, and treatments, including three IUIs. She tried acupuncture and changed her exercise routine, but nothing worked. "I was sad but not depressed. I would cry a lot when I talked to my husband or my mother about it. It was very emotional. But at the same time I really hated all that prodding and poking."

For Brenda and Janine, infertility was one of the worst experiences of their lives. But by joining my infertility program, by learning to relieve some of the stress of infertility, and by figuring out how to surround themselves with the love and support of family, friends, and other infertile women, Brenda and Janine conquered their infertility. That's what I'm going to help you do, whether you've been trying to get pregnant for six months or six years. With the dozens of mind/body techniques, coping strategies, and lifestyle changes in this book, I'll help you conquer infertility, too.

What Is Infertility?

The official definition of infertility is failing to produce a pregnancy that results in a live birth after one year of unprotected regular intercourse if you're under age thirty-five, and after six months if you're over thirty-five. Infertility is a major health problem in the United States, as these numbers show:

- Infertility affects 6.1 million women, or about 10 percent of the reproductive-age population, according to the American Society for Reproductive Medicine.
- Infertility affects men and women with almost equal frequency. According to Resolve, Inc., a national infertility support group with over fifty chapters throughout the United States, 35 percent of infertility is due to a female factor, 35 percent to a male factor, 20 percent is a combination of male and female factors, and about 10 percent is unexplained.
- Assisted-reproductive technology (ART) has been used in the United States since 1981. More than 70,000 babies have been born in the United States as a result of ART, including 45,000 as a result of in vitro fertilization (IVF).

Infertility is a growing problem in the United States, although experts don't know exactly why. One possible explanation is that more men and women are putting off childbirth until their thirties and forties, when aging eggs and sperm make conception more difficult. Environmental contaminants may also play a part.

The most common causes of male infertility are sperm-cell problems—some men produce few or no sperm cells, or the sperm they produce are either abnormal and/or can't swim effectively enough to reach the egg. Among women the most common problems are ovulation disorders and problems with fallopian tubes (often caused by scarring from pelvic inflammatory disease or endometriosis). Structural deformities of the uterus and uterine fibroids, which are also common among infertile women, may contribute to repeated miscarriages.

When you first see a doctor about your inability to get pregnant, there are several standard tests to assess the fertility of both you and your partner. You will be asked to have blood tests to assess your thyroid function, prolactin level, and other hormone levels. Tests for your FSH and estradiol levels will be done on day three of your menstrual cycle. You may be asked to keep a basal body-temperature chart or use ovulation-predictor kits to determine when and if you ovulate. Your partner will be asked to provide a semen sample. This should really be done after the first visit. (If this isn't done before you begin infertility treatment, think about getting a new doctor. I have had dozens of patients go through expensive and painful treatments before their partner was even tested, and the diagnosis turned out to be a male factor, for which the prescribed treatment could be different). You will likely be advised to undergo a hysterosalpingogram, in which dye is injected into your uterus to see if the uterus and tubes are normal. Other tests may be necessary, depending on what these initial tests show.

As for treatment, what your doctor recommends will depend on your diagnosis and may include medications such as clomiphene (Clomid), gonadotropins, thyroid-replacement hormones, or lupron. If you have endometriosis, adhesions, polyps, fibroids, or a uterine septum, or if your husband has a varicocele (a varicose vein in a testicle), surgery may be recommended. You might undergo intrauterine inseminations with or without medications. The most expensive and invasive procedure, but the one with the best success rates for many conditions, is in vitro fertilization. For more detailed review of infertility diagnostic tests and treatments, please refer to a book I wrote with two Harvard physicians, *Six Steps to Increased Fertility* (Barbieri, Domar, and Loughlin, published by Simon and Schuster/Fireside, 2001).

Although it takes six months to a year to receive an official diagnosis of infertility, many women—particularly those who are over thirty-five—begin to feel stressed and worried about their fertility shortly after they start trying to conceive. The stress builds as conception fails to happen the second month, the third, the fourth, the sixth. It grows as you start wondering if you'll ever get pregnant and skyrockets when you see doctors, undergo tests, endure invasive medical procedures, and still fail to get or remain pregnant. The stress builds and builds and, in my experience, hits its peak after two to three years of unsuccessful trying. For many women infertility is the most upsetting experience of their lives, a tragedy that causes as much stress as does a life-threatening disease, as it did for my patient Lorna.

Lorna, a thirty-six-year-old nurse, had been trying to get pregnant for about nine months. When nothing happened, her OB/GYN ran some tests. Lorna was devastated to discover that she had abnormal levels of follicle-stimulating hormone (FSH), a hormone secreted by the pituitary gland to stimulate growth of the small fluid-filled sac, or follicle, in the ovaries that nurtures a ripening egg. Her doctor also told her that her eggs had aged prematurely and were lower in quality than the average thirty-six-year-old's. The night that she received her test results, she spent hours on the Internet doing research on FSH and egg quality. The more she learned, the more frightened she became. "It dawned on me that I might never get pregnant," she says. "I estimated that I had only about a thirty percent chance of getting pregnant without a donor egg."

Lorna immediately began to feel depressed, and her symptoms grew as she underwent injections of fertility drugs, three IUI procedures, and one in vitro fertilization, all of which failed. "I was really in a bad way," Lorna recalls. "I wasn't sleeping, I wasn't eating, and I felt nauseated. I'm small to begin with, and I lost ten pounds. It was terrible." She would wake up at 3:00 A.M. and lie awake for hours, sometimes in a panic over her treatment. "Because I'm a medical person, I would obsess about medical details—I would think, 'Oh, they didn't put me on enough medication, I know I'm not going to make enough follicles. They really should have done this or that differently'—which is not wise but hard to avoid." Because of her dark moods and sleeplessness, she had trouble concentrating at work and felt anger toward the pregnant patients that she had no choice but to care for in the medical office and hospital in which she worked. Her husband tried to help her, but she was inconsolable, and he felt tremendously frustrated because he could do nothing to alleviate her emotional anguish.

Infertility affects every aspect of a woman's life, from her relationship with her family and friends to her career. Often the first thing to be affected is her relationship with her husband, and for many couples infertility is the first crisis—the first real test—of their marriage. Most couples I counsel say that, looking back after their infertility is resolved, they realize that the experience brought them closer together, that they forged a deep bond during the process. While they're in the middle of the infertility crisis, however, husbands and wives often find themselves out of step with each other. Although many couples agree that they both want to have children, I have never in my career seen a couple who are in the same place at the same time regarding pregnancy and infertility. The women usually, although not always, seem to be about a year ahead of the guys. The woman thinks there's a problem before the man does. She wants to move on to medical procedures such as IVF before he does, and she is prepared to think about adoption before he is. Often, when I meet privately with infertile couples, the woman says, "He's holding me back," and the husband says, "She's pushing me to be ready for things I'm not ready for." This can cause tremendous conflict within a couple.

Part of the problem is that men and women respond to infertility differently. Women tend to express a lot more distress than men do. In fact, in a study of two hundred couples being evaluated for acceptance into an IVF program, 48 percent of the women, in contrast to 15 percent of the men, reported that infertility was the most upsetting experience of their lives. That leads to situations like this: The woman gets her period and is crying hysterically, and the man is stoic, saying, "Don't worry, you'll get pregnant next month." She feels he doesn't care, and he feels as though she's going off the deep end. He thinks that only a baby will make her happy, and that makes him feel as though he's not enough for her. After a while, both the man and the woman start to question why they married this person. They each ask themselves, "Why isn't he [or she] reacting the same way that I am?" The woman wants the husband to be more upset; the husband wants the wife to be more rational. They fight about it. They withdraw from each other. They question whether their marriage would even be strong enough for a baby when they're being torn apart by the conception process.

Of course, all this stress and conflict spills over into your sex life. When you're trying to conceive—particularly when you're undergoing medical treatments—you're told when you can and can't have sex. Your sex life is directed by a doctor or a calendar or an ovulation kit, not by the passion and love you feel. Eventually men may start to feel as though their wives want to make love with them only to extract sperm from their bodies—particularly if their wives lose interest in sex completely and only want to make love midcycle, when they're fertile. In fact, a lot of women begin to associate sex with failure and avoid it whenever possible. Infertility robs you of the sensuality and spontaneity of making love, and sex becomes solely the mechanical action of trying to make a baby. The only time many men feel comfortable being intimate with their wives is in a sexual situation, so the loss of loving, intimate sex can have a devastating impact on the marriage, as both husband and wife pull away from each other. Guilt and blame may follow, pushing you even further apart.

Spreading Stress

Infertility can impair the relationship that couples have with their families, too. You may feel jealous of siblings who conceive and deliver healthy babies with ease. I've seen many women who have never before experienced an envious moment in their lives become very jealous when someone else gets pregnant. It's especially hard when a younger sister conceives, and you find yourself filled with the kind of sibling rivalry you haven't felt since you were kids.

Parents often aggravate the situation by repeatedly asking when they will become grandparents or by talking endlessly about their other grandchildren. Or they may tell the couple that if they'd just relax, take a vacation, and calm down, they'd get pregnant. Some unenlightened family members blame infertility on the woman's career, urging her to quit her job or take a demotion to a less stressful position. If the couple chooses not to share information about their infertility with their families, their families may assume that the couple isn't even trying to get pregnant so that they can selfishly enjoy a double-income, no-kids existence and deprive their parents of the grandchildren they crave. When couples feel such a lack of understanding among their families, they are reluctant to go home for holidays. They dread being surrounded by their nieces and nephews and pregnant siblings and nagging parents. So they make their excuses, drift away from their families, and lose potential support.

Relationships with friends can suffer, too, particularly if the couple's friends are getting pregnant easily. Few things are more difficult for a woman struggling with infertility than being surrounded by pregnant women and babies. So infertile women begin to avoid their friends and eventually start to sacrifice their social network at a time that they need the support of loving friends more than ever.

Infertility intrudes at work, too. Medical treatments can limit both the husband's and the wife's ability to travel for work and to schedule appointments. Infertility treatments require that the woman make frequent doctor visits, have blood work, ultrasounds, and early-morning clinic appointments that play havoc with a work schedule. Husbands often have to be available midcycle to provide sperm, and if a woman needs daily injections, she might need her husband to administer them. I actually had a patient who was fired from her job because she arrived late to work repeatedly during an IVF cycle—each day she had to be at the hospital at 8:00 A.M., but she didn't want to tell her employer why she was late because she was afraid she'd be discriminated against. I've also had a lot of patients who haven't taken promotions because their new positions would require more travel. Some of my patients have even quit their jobs because they found that dealing with infertility was a full-time job in itself, and they couldn't successfully juggle the demands of their career with the stresses of infertility. It can also be difficult when co-workers get pregnant—I remember that one of my patients had four co-workers who were pregnant at the same time, and they talked about babies constantly. My patient couldn't escape from these chatty colleagues. After all, you can avoid your friends, you can even avoid your family, but unless you quit your job, it's nearly impossible to avoid the people you work with. Going to work every day was extremely painful for this woman and a tremendous cause of anxiety.

Infertility can lead to an intense spiritual crisis. A lot of my patients find infertility to be the first time their prayers are not answered by God, and they lose faith. They think that God is punishing them or angry with them, or that God feels they would be poor parents. Many begin to doubt the existence of God, and they forfeit yet another source of emotional support. Catholics may feel even more tension, because the Catholic Church prohibits some forms of assisted-reproduction treatments and they must choose between disobeying their church's teachings and giving up their dreams of being parents. The vast majority of Catholics do go ahead with infertility treatments despite their church's objections, but I've had a number of patients who have really struggled with that decision, and they ultimately (and often bitterly) decide that they cannot go against religions teachings. Others make peace with their decision only after a liberal priest tells them, off the record, to go ahead with infertility treatments because procreation is so important to the Catholic Church.

And then there's money. Infertility treatment is very expensive, and there's no guarantee that it will work. For example, IVF costs somewhere around ten thousand to twelve thousand dollars per cycle but is successful only 25 percent of the time. If your health insurance doesn't cover such treatments—and at last count some thirty-nine states have no mandated coverage of infertility treatments—infertility can carry a tremendous financial burden. Often couples borrow huge sums of money to cover treatment, and if it fails, the couple may have no resources left to pay for adoption. Financial problems can cause arguments, particularly when the husband and wife disagree on how many rounds of treatment they want to try or whether they want to adopt.

 How Stress Harms

The psychological distress of infertility can do more than affect your marriage, your relationship with others, and your career. It can take a tremendous toll on your health. Researchers estimate that some 80 percent of chronic health problems—including heart disease, cancer, and hypertension—are exacerbated by stress. And, ironically, stress can make you even less likely to conceive. Both my clinical experience and the research indicate that women who are highly stressed and depressed are less likely to become pregnant, either naturally or via assisted-reproductive techniques.

If you're struggling with infertility, stress creates a vicious circle: You get stressed because you can't conceive, which makes you more stressed, and that makes it even harder for you to conceive. And that stresses you out more! This process can spiral on endlessly unless you learn to break the cycle of stress. I'm going to teach you how to break that cycle. But first I want to share with you some of the things I have learned about infertility, stress, and depression.

Everyone knows what it's like to be stressed out. Think of how you felt the week before your wedding. You were running around from store to store, having last-minute adjustments on your wedding gown, calling the caterer, arguing with the travel agent, preparing for visitors, cleaning the house, staying up late, getting up early, eating junk or skipping meals, neglecting your exercise, and focusing completely on a future moment—your wedding day—rather than on the moment you were in. And it wore you out so much that on your honeymoon all you wanted to do was order a blender drink, lie down on the sand, and destress for a week. But stress does more than just wear you out. Yes, it leaves you feeling fatigued, overwhelmed, and emotionally burned out. But it also causes tremendous changes in your body's biochemistry and rhythms. It upsets your body's balance in a very far-reaching way, and over time it can lead to or worsen chronic health problems.

When something upsetting happens to you, your body undergoes a very real and very dramatic physical response. Say you're driving along a country highway, calmly listening to some music on the radio, when a deer suddenly runs out onto the road in front of you. In an instant your body responds. Your heart begins to pump like crazy, rushing oxygen-rich blood to your muscles. Your blood pressure skyrockets, and your breathing becomes shallow and rapid. Stress hormones such as adrenaline, noradrenaline, and adrenocorticotropic hormone pour into your bloodstream. Your muscles tense up, your immune system and digestive system temporarily shut down, your brain switches into a state of hyperalertness, and your hypothalamus releases endorphins, the body's natural painkillers.

Nearly every system of your body becomes immediately prepared to help you deal with the stressful situation. This response is referred to as "fight or flight" in honor of our ancient ancestors, who, when faced with something ferocious, would either have to fight with it or run away from it in order to avoid being gobbled up by it. Likewise, that same physical response gives you the wherewithal to slam on your brakes or swerve out of the way to avoid hitting the deer and being seriously injured yourself.

Let's look at what happens next. You've safely managed to avoid hitting the deer. As the deer darts off into the woods, you swear profusely at it, shaking your fist and shouting that it deserves nothing better than to be boiled up as venison stew by the nearest hunter. Then you take a deep breath, thank your lucky stars that you were able to stop in time, check to make sure that you and the car are both okay, and then go on your way. As you drive off and leave the stressful situation behind, your systems gradually return to normal. Over time your heart rate, blood pressure, breathing, stress-hormones, immune system, digestive system, and mental clarity all revert to their ordinary levels. Eventually you start humming to the music on the radio again, and you return to a calm, relaxed state.

But what would happen if another deer ran out in front of your car forty-five minutes later? And then again, an hour after that? What if you were repeatedly subjected to fight-or-flight distress, bombarded by alarming events hour after hour, day after day? Your body would react to such relentless threats by remaining in a constant heightened level of response, and it would never have a chance to return to normal. Your heart rate, breathing rate, and muscle tension would stay elevated. Your body would overflow with stress hormones. Your immune system would remain suppressed, compromising your ability to fight disease. Constant surges in blood pressure and cholesterol production could damage blood vessels. And frequent demand for endorphins would actually deplete their power, aggravating migraine headaches, backaches, and other chronic ailments that are soothed by normal endorphin secretion. This assault on your body could contribute to insomnia, exhaustion, increased production of stomach acid, irritability, sexual dysfunction, and a host of other ailments.

This is exactly what can happen to infertile women. Instead of car accidents and darting deer, infertile women are confronted by a nonstop barrage of negative test results, hurtful self-talk, anxiety, guilt, unwanted menstrual periods, and stressful social interactions. Their bodies respond to the relentless stress of infertility in the exact same way that they would if a deer ran out in front of their speeding car several times a day: by staying in red-alert mode almost constantly.

Which Comes First, Infertility or Stress?

We know that infertility causes stress, but can stress cause infertility? That depends, in my opinion, on how you define stress. Until recently most researchers interpreted stress as anxiety and used measures of anxiety in their search for a connection between stress and infertility. (Some studies suggest that anxiety does influence infertility, and others have failed to support such a relationship.) However, if you define stress—as I do—as a combination of the physiologically damaging depression, anxiety, and isolation that infertile women relentlessly feel when they fail to get or stay pregnant year after year after year, then, yes, I do believe that stress can cause or exacerbate infertility. And research from around the world and my clinical experience support this thesis.

Let's back up a bit. The question of whether stress causes infertility has been bandied about for thousands of years, ever since Hannah, in the Bible's first book of Samuel, wept, wouldn't eat, and carried grief in her heart because "the Lord had shut up her womb." More recently the medical and mental-health communities have pushed the infertility cause-and-effect pendulum back and forth, sometimes arguing that infertility is caused by stress and anxiety, sometimes arguing that emotional factors have nothing whatsoever to do with infertility. In the 1940s, 1950s, and 1960s, the psychiatric community began to claim that women were barren because they were conflicted about motherhood. But those theories were shot down when researchers discovered that infertile women were no more conflicted about motherhood than were their fertile sisters. Then, as medical technology improved and doctors became better able to diagnose and recognize tubal blockages, ovulation problems, early menopause, and male-factor issues, the pendulum swung in the opposite direction, and researchers began to believe that all aspects of infertility were physiological and that the psyche had nothing to do with it. In the 1980s and early 1990s, this belief was nearly universally confirmed, because studies looking at anxiety showed that anxiety was not a deterrent to pregnancy when high-tech interventions were used—meaning that anxious women who used assisted-reproductive technology seemed to be as likely to get pregnant as those who weren't anxious. These studies convinced many in the medical establishment that infertility was physiological and had no psychological component.

But those studies looked at anxiety, not depression. Recent studies that consider depression suggest that it plays a strong part in infertility—for example, one study found that women with a history of depressive symptoms were nearly twice as likely to report subsequent infertility than were women who were not depressed. Another study determined that women who had experienced at least one unsuccessful IVF cycle and who had depressive symptoms before continuing IVF treatment experienced a 13 percent subsequent pregnancy rate, in contrast to a 29 percent pregnancy rate in women who did not experience depressive symptoms before their IVF cycle. And there's one more study I should mention—I promise, I won't talk about studies much longer, but this really is important: In this study, women with female-factor infertility who had increased depressive symptoms on day three of their IVF cycle experienced significantly lower pregnancy rates than did women who were not depressed.

That is some pretty strong evidence, in my mind, that depression can contribute to infertility. And really, the results of those studies aren't a bit surprising to me, based on what I've seen among the women in my infertility groups. I'm convinced that depression plays a part in infertility, and that reducing depressive symptoms generated during the grueling process of infertility treatments can increase the chance of pregnancy.

Wait a minute, though. We know that women are at high risk of depression after they've been on the infertility treadmill for a while. What about the woman who is just starting to try to get pregnant? Can depression affect her fertility?

Yes, I believe that it can. Infertility certainly isn't the only thing that can trigger depression—plenty of women without a history of infertility suffer from depression, including many who have mild or moderate depression and don't even realize it. If you are depressed before you start trying to conceive—even mildly depressed—your fertility may be compromised, and you might not conceive right away. Add on a factor such as a elevated FSH (which may be an indication of poor egg quality or diminished ovarian reserve) and your chances of conceiving quickly are even lower. Then, if you don't get pregnant right away, your depression may deepen rapidly, making conception even more elusive. Considering that even the most fertile couples have only a 15–20 percent chance of conceiving during any given month, and considering also that the strain and anxiety that come from not being able to conceive begin, for many women, after just a couple of negative pregnancy tests, it's clear to me that depression can cause some measure of infertility in women who are reproductively sensitive to it.

 The Depression Connection

The idea that depression can limit a woman's chances of conception is a relatively recent one, but it's something in which my colleagues and I have long had an interest.

In 1986 cardiologist Herbert Benson, my mentor and the founder of Harvard's Mind/Body Medical Institute, gave a talk about stress and the relaxation response during OB/GYN grand rounds at Beth Israel Medical Center in Boston. (In the late 1960s Dr. Benson was the first researcher to discover and give a name to the phenomenon known as "the relaxation response." This is a physical state of deep rest that occurs when a person is deeply relaxed. It interrupts the body's physical and emotional responses to stress, allowing it to return to a calm, relaxed state. When a person elicits the relaxation response, there is a measurable decrease in heart rate, blood pressure, breathing rate, stress-hormone levels, and muscle tension. Elicitation of the relaxation response is at the heart of our mind/body programs.) During his talk, Dr. Benson was discussing the physiology of the relaxation response, and he mentioned that it is mediated by the hypothalamus. Dr. Machelle Seibel, who was then the head of the Beth Israel infertility program, posed a question: "Because the hypothalamus regulates all aspects of reproduction, couldn't there be some relationship between the relaxation response and infertility?" And Dr. Benson and I thought, "Hmmm, here's a research project that looks interesting."

So we decided to do a study. We were going to recruit a hundred women with unexplained infertility. Half would be trained in eliciting the relaxation response, and half would be controls—they would not learn relaxation techniques. We would follow them for six months and watch for the differences in pregnancy rates and psychological status. We began recruiting women, and, as it happened, the first three women recruited into the study were randomized into the control group, which means they wouldn't receive any relaxation training at all. They were not happy about this. In fact, I couldn't get them to stop crying long enough to fill out my forms. They had the expectation that I'd be helping them, and when they heard that they were in the control group, they began to despair.

Meanwhile, I was helping to run mind/body groups for people with other medical conditions. What I saw astounded me. People with lupus and multiple sclerosis and migraines and other medical problems were participating in ten-week mind/body programs, and within just a few weeks they were starting to feel better.

I had come into the mind/body program from a very scientific Ph.D. program, and therefore was very science oriented—so I was skeptical. But there was no denying what I was seeing: People were coming into these programs with organic disease and having their symptoms reduced through a mind/body approach. I was shocked. I knew what I needed to do. One minute I was watching people with serious diseases feel better thanks to mind/body strategies, and the next minute I was listening to infertility patients sobbing because they were going to be controls in my study. I had to help these women. I went to Herb Benson and said, "Look, I can't keep on randomizing these women into control groups. Let's postpone the study and start a clinical mind/body program for infertility instead." He agreed. So in September 1987 we started our infertility program. I'd guess it was the first time in Harvard Medical School history that a study was scrapped because the subjects cried.

We began our mind/body program for infertile women on little more than a hunch. Although we had solid evidence that mind/body strategies helped patients with a range of other diseases and ailments, we had very little evidence that our program would work for infertility patients. A very small, unpublished 1985 study out of Bogotá, Colombia, had looked at a group of women with unexplained infertility. Some of the women received mind/body therapy, and a control group didn't. By the end of the study, four of the seven in the mind/body group got pregnant, and none of the control group did. This was the first and only study to show such a connection, and it hadn't even been published. But we also had a strong feeling that even if we couldn't help these women to conceive, we definitely could make a difference in the quality of their lives. They were clearly showing symptoms of depression, and we were sure we could help reduce their depression.

We were right. The infertile women who took part in that first ten-week program felt better. And they felt better fast. Their depression lifted, their anxiety levels lifted, their anger lifted. Every psychological parameter we looked at got better very quickly, and about a third of the patients became pregnant within six months. (This compares to a six-month pregnancy rate of 18 percent in a Canadian study of women receiving infertility treatment but no mind/body treatment.) Those pregnancy rates have held up over the years. Keep in mind that the patients in our infertility program are women who have tried all the preliminary infertility treatments—they've already tried all the easy stuff. When they get to us, most have completed an infertility medical workup, half are doing IVF, and the other half are thinking of it. Virtually all of them have received an official diagnosis of infertility from an OB/GYN or reproductive endocrinologist. Our patients are older women, women with borderline FSH levels, women who have had recurrent miscarriages. Women who are going to get pregnant easily don't come to us. Our average patient has been trying to conceive for more than three years.

Finding Answers: Depression Research

About a year after we started our infertility program, the National Institute of Mental Health (NIMH) put out a bulletin saying it wanted to fund depression research. I flew down to Washington, D.C., to tell the NIMH that I had this great program that treats depression in infertility patients and to ask for some money to expand my research. The NIMH queried, "How do you know these patients are depressed?" And I said, "Just ask them. They cry all the time; of course they're depressed." But that wasn't enough for the NIMH. They wanted hard data, so I went back and got it. We collected data on 338 infertile women and 39 fertile controls (women who were coming in for routine annual Pap tests). We found that indeed the infertile women were more depressed than the fertile women. About a third of all infertile women were depressed, compared with about 18 percent of fertile women. Another study showed that 11 percent of infertile women met the criteria for having had a major depressive episode, compared to 3.6

percent of fertile women. We also found that it was almost impossible to predict who in the infertile group would be more depressed. Whether or not they already had a child didn't make a difference, their age didn't make a difference, their diagnosis didn't make a difference, whether they'd had a pregnancy loss didn't make a difference. The only thing that was strongly associated with depression was how long the women had been trying to get pregnant. The most depressed women had been trying for two to three years—and that makes sense clinically. The first year you're trying on your own. The second year you start seeing a physician, and you're optimistic that he or she will help you. And after a year of that, you begin to think nothing will help. That's depressing.

I went back to the NIMH with my depression data. And the NIMH said, "How do you know it's the infertility that is depressing them?" At this point *I* got depressed. It seemed so obvious to me that the women were depressed because of their infertility, but again the NIMH wanted hard data. So again I gave it to them.

All of our patients receive psychological testing as a part of the mind/body program. (We utilize a common psychological screening tool that measures depression, anxiety, anger, and so on.) We compare infertility patients' psychological status before and after the program. We also compare the psychological scores of the infertility patients with the patients in other mind/body programs who have heart disease, cancer, HIV-positive status, and chronic pain. The first time we did this, we were stunned at the results: The infertility patients were just as depressed and anxious as all the other women except the chronic-pain patients, who were the most depressed. This was an astounding finding—*the infertile women were every bit as depressed as the people who were confronting illnesses that could kill them!* So I went back to the NIMH and showed them three things:

1. Infertile women are more depressed than fertile women are.
2. Their depression levels peak two to three years after they start trying to conceive.
3. Infertility has as great a psychological impact as does a potentially terminal illness.

Finally, in 1994, the NIMH funded a study. During the years that followed, we kept careful track of the emotional states and pregnancy rates of different groups of infertile women before and after mind/body intervention. In 1999 and 2000 we published two studies that demonstrated the success of our program. I know, here I go again with the studies, but I promise to make this brief.

In a study my colleagues and I published in the *Journal of the American Medical Women's Association* in 1999, we analyzed the data collected from 132 infertile women who attended our mind/body program. We found that 42 percent of the 132 infertile women who participated in our program conceived within six months of completing the program. Not only that, but the more depressed a woman was before the program started, the more likely she was to get pregnant after participating in the program. This is a dramatic finding, because it gave us solid evidence that depressive symptoms can hamper a woman's fertility. **And it also told us that women could play an active role in reducing their depression by using the mind/body strategies we teach in our mind/body infertility program.**

The next year we published an article in the journal *Fertility and Sterility*. In that study we looked at 184 women who had been trying to get pregnant for one to two years. In our study we placed women in one of three groups: a mind/body group that learned many, but not all, of the techniques taught in our program; a support group that met weekly to discuss the impact of infertility; and a control group, which received no intervention at all. The results were dramatic: **Within a year 55 percent of the mind/body and 54 percent of the support-group participants conceived pregnancies that resulted in a baby, compared with only 20 percent of the control group.** Not only that, but according to a study we published in the journal *Health Psychology* in 2000, the women in the mind/body group were significantly less depressed six months later than were the women in the support group and the control group. And although the numbers were not statistically significant, more women in the mind/body group (42 percent) conceived their babies naturally, without medical intervention, than did those in the control group (20 percent) and the support group (11 percent).

The bottom line? **Alleviating depression and other psychological**

distress in infertile women appears to make it easier for them to become pregnant.

Why would depression make a woman less fertile? The medical world doesn't know for sure. Depression may reduce egg quality, delay the release of eggs, prevent the implantation of a fertilized egg, or decrease levels of reproductive hormones necessary for a fertilized egg to grow and thrive. Depression may also destabilize a woman's hormone levels: One study suggested that depression is associated with abnormal regulation of luteinizing hormone, which plays a major role in conception. Or maybe the immune system, which can be weakened by stress and depression, interferes with conception in some way. We just don't know for sure. But here's what I think: When you look at it from an evolutionary point of view, it makes sense that depressed women would have trouble conceiving. If a woman is clinically depressed, she is unable to care for herself, let alone a baby. Infertility may be nature's way of postponing pregnancy until the mother is psychologically capable of raising a child. We've seen that animals who are in hopeless situations are less fertile, so it is not at all surprising to me that the same might be true of people.

Stress and depression affect men, too. Studies show that men with previously normal sperm counts are eight times more likely to have low sperm counts after a year or two of infertility. And a study in Germany of men on death row—you can be pretty sure those guys were stressed and depressed—found that every one of them had a low sperm count.

I'm not saying that depression affects everyone's ability to procreate. After all, women have been known to get pregnant easily and spontaneously while being hospitalized for severe depression in a psychiatric hospital. But some people are reproductively sensitive to depression, and those are the people who can benefit from either medication or mind/body strategies to reduce depression and stress and improve the odds of conceiving. As for anxiety, the research points to the conclusion that anxiety doesn't cause infertility. In my clinical experience, however, anxiety usually comes hand in hand with depression. It covers up depression. When doctors treat a person for anxiety, they often discover, after clearing up the anxiety, an underlying depression that must be treated separately.

 Are You Depressed?

Depression is a debilitating illness that affects you physically, emotionally, mentally, and socially. It can disrupt every part of your life. Although some people are genetically predisposed to depression, it can also be triggered by conditions of extreme stress or grief—such as those brought on by infertility—even in a person with no depression in her own past or her family history.

Depression is a disease with biochemical origins. It is not a weakness of character, it is not something that happens to people who don't try hard enough to be cheerful, and it is not something to feel guilty about or ashamed of. *If you are depressed, it is not your fault.* I cannot stress this enough. Many people, especially women, think that the emotional symptoms caused by depression aren't legitimate and that when they are depressed, they should just be able to "snap out of it" through sheer willpower. That is just not true. Depression is a very real, biologically based disease that needs to be treated with medication, psychotherapy, or both.

If you have been unable to conceive a baby for more than a few months, I urge you to consider very seriously whether you may be depressed, even if you think you're handling it just fine. One of the hallmarks of depression is that depressed people often do not recognize or admit that they are depressed. I can't tell you how many of my patients have sat in my office telling me, "I'm not depressed. Really! I'm just a little blue!" only to receive a diagnosis of moderate or even severe depression a week later from a psychiatrist. Remember Lorna, the patient I mentioned earlier in this chapter whose infertility left her feeling so miserable and overwhelmed? After a huge amount of encouragement from her husband, Lorna finally saw a psychiatrist, who diagnosed major depression. Major depression! And she didn't think she was depressed. Even though she felt crummy all the time and had trouble dragging herself out of bed some mornings, Lorna was still amazed at that diagnosis—and remember, Lorna is a nurse, so she's more aware of such things than the average woman on the street. "I really didn't think I was clinically depressed," Lorna says. "But when my psychiatrist told me that I was, it was sort of like, 'Oh, yeah, I guess you're right.' And when I looked back on it months later, it was so clear to me how very depressed I was. But I just didn't realize it then."

If you think you might be depressed, the following checklist from the American Society for Reproductive Medicine can be a helpful self-diagnosis tool. If you experience any of the following symptoms over a period of two weeks or longer, seriously consider talking with a mental-health professional as soon as possible. Patients with fewer than five of these symptoms are said to have minor depression; those with more than five symptoms are diagnosed with major depression:

- Loss of interest in usual activities
- Depression that doesn't lift
- Agitation and anxiety
- Marital discord
- Strained interpersonal relationships with your partner, friends, family, or colleagues
- Difficulty thinking of anything other than infertility
- High levels of anxiety
- Diminished ability to accomplish tasks
- Difficulty concentrating
- Change in sleep patterns (difficulty falling asleep, staying asleep, or waking up early in the morning)
- Change in appetite or weight
- Increased use of drugs or alcohol
- Thoughts about death or suicide
- Social isolation
- Persistent feelings of pessimism, guilt, or worthlessness
- Persistent feelings of bitterness or anger

Depression should be treated, not just because it may be hampering your ability to get pregnant but because if it is left untreated, it can lead to long-term health problems.

For example, recent research shows that people with depression are four times more likely to suffer a heart attack than are those who are not depressed. Researchers believe that part of the reason for this increased risk of heart attack is that depression may result in chronically elevated levels of stress hormones such as cortisol and adrenaline, two of the hormones released during the body's fight-or-flight response. Chronically high levels of these hormones can hurt the heart. In addition, many of the body's physiological reactions to depression and distress—rapid heartbeat, high blood pressure, faster blood clotting, elevated insulin and cholesterol levels—can damage the heart and the cardiovascular system. They also divert the body's metabolism from routine tissue repair and other everyday maintenance functions that keep the heart healthy over the long term.

Depression may also contribute to the bone loss that causes osteoporosis, broken bones, and hip fracture. Recent research suggests that excess cortisol secretion, a common feature of some forms of depression, can cause irreversible bone loss.

Even mild depression can hurt you. Another recent study found that persistent mild depression may lower immunity and the ability to fight off disease. The researchers suggest that depression may compromise a person's ability to fight off infections or even cancer.

Even if you are not clinically depressed, you (and your husband, perhaps) may benefit from meeting with an infertility counselor. An infertility counselor is a psychologist, psychiatrist, social worker, psychiatric nurse, marriage counselor, family counselor, or clergy member who is specially trained in infertility-related issues and who can help you think through your feelings, develop coping mechanisms, and make decisions related to infertility treatment. Many infertility clinics have such counselors on staff and may even require that you meet with them before undergoing certain procedures, such as IVF or using donor eggs or sperm. To find an infertility counselor in your area, look at the American Society for Reproductive Medicine Web site (www.asrm.org) for a list of mental-health professionals, ask your OB/GYN or reproductive endocrinologist for a referral, or contact Resolve. Resolve's contact information is listed in the Resources appendix in the back of this book.

Resolve can also provide information on how to find an infertility support group. Support-group participants meet regularly to talk about the challenges of infertility and to share solutions, coping strategies, medical information, and support. I'll talk more about the many benefits of group support in the next chapter.

 What About Antidepressants?

If you are depressed, your doctor may recommend antidepressant medication. But is medication the right choice for you? Will it help you become pregnant? If you take medication and become pregnant, will it pose any risk to the fetus?

Those are complicated questions. Some studies show that antidepressant medication can boost fertility. A study in India, for example, found that infertile women who took psychiatric medications were more likely to become pregnant. And clinically, I've seen quite a few depressed women in my program get pregnant after taking Prozac or other antidepressant medications. But the problem is, we have no definitive proof that antidepressants are safe for a fetus. Preliminary research suggests that women who take Prozac appear to have healthy children, but not enough time has elapsed for us to know this for sure. Drug companies may be reluctant to conduct research on the safety and efficacy on women who are trying to get pregnant because of potential liability issues, and I don't blame them—nobody wants to risk putting unborn babies in danger. So it may be some time before we settle this question.

For women with major depression, however, the benefits of medication can outweigh possible risks. If you think you are severely depressed, or if people you love are telling you that *they* think you are very depressed, put this book down right now and make an appointment with a psychiatrist. *Severe depression must be treated.* If you're worried about the side effects of medication on a fetus, consider taking a break from conception efforts while you're on antidepressants, and then resume your efforts after you've stabilized your mood and have stopped using the medication. (Taking a break from trying to get pregnant can be an effective strategy even if you aren't on antidepressants.) Or take the antidepressants only until you know you have conceived, and then discontinue them with your doctor's approval.

In many cases of mild or moderate depression, mind/body therapy can be as effective as an antidepressant. However, you need to see a psychologist or psychiatrist to assess your symptoms. If you are experiencing minor depression, it may be treated effectively with mind/body therapy. (In the next chapter I'll teach you everything you need to know about how to use those strategies.) But if you are experiencing a major depression, medication, at least temporarily, would be the most efficacious treatment.

Now, you're probably wondering how things worked out for my patients Brenda, Janine, and Lorna. Here's how their stories end:

Brenda's Story

Noticing that Brenda needed emotional help, her doctor recommended my mind/body infertility program. She waited several months to sign up, but eventually she decided to give it a try. Within weeks of beginning the program, Brenda was feeling better. "Going to those meetings was something I did just for myself," Brenda says. "I realized that I hadn't been taking very good care of myself." Using relaxation techniques, mindfulness, and a number of other strategies that I'll describe in Chapter 2, Brenda began to regain control of her emotions. She became less depressed, her marriage grew stronger, and she felt better prepared to cope with whatever her future might hold.

Brenda became pregnant before the end of the ten-week program, and she delivered a son the following summer. Everything worked out well for Brenda, but she does have some regrets. "I wish I hadn't wasted a year of my life feeling so miserable. I should have seen a therapist or signed up for the mind/body program sooner. I should have just taken care of myself better, and gotten myself fixed emotionally, and not spent a year so unhappily. I think if I hadn't been so emotionally devastated, I would have gotten pregnant. I think I was just too miserable to get pregnant."

Janine's Story

Unsure of what to do next, Janine signed up for my program. Should she try IVF? Donor eggs? A surrogate? Adoption? Janine couldn't decide. Although she wanted to be a mother, she felt reluctant to go to such extraordinary means to have a child, and she was concerned about what years of infertility treatment might do to her marriage. "I became very protective of my own self, of my own soul, of my new marriage. I found the procedures more emotionally invasive than physically invasive." Janine discovered that talking in depth with the other infertile women in our group helped her enormously. "I think I learned the most by listening to other women," Janine says. "I formulated a lot of my viewpoints by observing and really seeing how all of the treatments affected everyone else's lives. It really shaped the perspective that I gained for myself."

Janine decided, with the support and encouragement of her husband and the women in the group, to stop trying to get pregnant and to choose not to adopt. "It felt like the right decision for so many reasons," Janine says. Today Janine has put her infertility behind her, is living a happy, childless life, and focuses her maternal energy elsewhere in her life, such as her career and her relationship with her husband. "I'm at peace with my decision," she says.

Lorna's Story

Lorna's psychiatrist wrote a prescription for a low daily dose—half the normal adult dosage—of Zoloft, an antidepressant medication. Lorna held on to the prescription for a week, unsure of whether to take it, but she finally decided that it was the right choice for her, and she had the prescription filled. She also started seeing a therapist regularly, and at about the same time she signed up for my mind/body program.

Within a few weeks she started to feel better. "I started to do relaxation exercises and take care of myself. For the first time I felt more hopeful than anxious. My husband said he was so happy to have his wife back." Toward the end of the ten-week mind/body program, Lorna had her second IVF—and this time it was successful. She went off the antidepressants as soon as she received news of her positive pregnancy test.

Today Lorna has a seventeen-month-old boy, her depression is a thing of the past, and she's planning to try for a second child. She continues to practice the mind/body strategies that she learned in class and has found that they help her to cope with other difficult situations in her life as well. "I have no doubt that they made a difference," Lorna says. "They are life skills, and I'll use them forever." And although Lorna would like to have a second child, she has come to a peaceful acceptance of the fact that future attempts to conceive may fail. "I'd love to have another baby," she says. "But it won't be the end of the world if I don't."

CHAPTER TWO

A Toolbox Full of Coping Skills

I keep a clever little tool in my kitchen drawer. It's a small hammer with a screw-off handle. Inside the handle is a handy-dandy two-sided screwdriver. One side of the screwdriver has a standard slotted head, and the other is a Phillips-head. This is great because, although it fits easily into the drawer where I keep my spatulas and wooden spoons, it takes care of just about any job that comes along in my kitchen: hammering a nail to hang a picture, unscrewing the back of a radio to insert batteries, tightening a loose drawer knob, even prying open a can of paint.

My little hammer is a perfect tool for me, but think of how useless it would be to someone who is building a house. A professional builder would need half a dozen different kinds of hammers and a whole array of screwdrivers, from a tiny slotted-head to an enormous power-driven Phillips-head screwdriver. Not only that, but a builder would need a truckload of other tools—wrenches, pliers, nail guns, saws, and others that I can't name and probably wouldn't even recognize.

The same is true with coping skills. When life is sunny and problems are few, you need only the little kitchen-hammer version of coping skills. If some bad news comes along, you deal with it by complaining to your best friend, or asking your husband for advice, or going out to lunch with your mother, and in a day or two everything is happy and sunny again. But when you've got to deal with a huge stressor such as infertility, which you may consider to be the worst thing that's ever happened to you, those kitchen-hammer coping skills just aren't going to do you much good. You need a huge toolbox full of coping skills, more than you may ever have needed before. And that's what I'm about to give you here.

As we discussed in Chapter 1, the research that my colleagues and I have done shows that alleviating depressive symptoms and other forms of distress appears to make it more likely for women to become pregnant. What I'm going to do is show you how to do that for yourself, both to increase your chances of getting pregnant and to help you feel better by regaining control of your life. In this chapter and throughout this book, I will share the many techniques that I teach the women who participate in my mind/body infertility program.

As I mentioned earlier, many of the techniques I'll teach you are designed to elicit the "relaxation response," our inborn capacity to reduce internal stress. Because eliciting the relaxation response is such a fundamental part of our mind/body program, I'll define it for you again:

> **The relaxation response is a physical state of deep rest that occurs when a person is extremely relaxed. It counteracts the body's physical and emotional responses to stress, allowing it to return to a calm, relaxed state. When a person elicits the relaxation response, there is a measurable decrease in heart rate, blood pressure, breathing rate, stress hormone levels, and muscle tension.**

Relaxation techniques cool down your nervous system and allow heart rate, breathing rate, muscle tension, and oxygen consumption to fall below normal resting levels. The relaxation response also causes normal waking brain-wave patterns to shift to predominantly slower patterns and pave the way for the hormonal and immune systems to return to normal. In some people, blood pressure decreases.

Because of these physiological changes, regular elicitation of the relaxation response helps lighten the symptoms of a host of health problems, including insomnia, chronic pain, hypertension, the side effects of cancer treatments, menopausal hot flashes, irritable bowel, migraine, anxiety disorders, PMS symptoms, and eating disorders.

Relaxation has long-term benefits as well. Researchers at the Mind/Body Medical Institute were the first to show that people who consistently elicit the relaxation response for two to six weeks begin to notice a "carryover" effect: Rather than feeling better for just minutes or hours after their practice, patients begin to feel better twenty-four hours a day. The calm, peaceful feeling of relaxation carries over into their everyday life. That's not a bad payoff for an investment of twenty minutes a day, is it?

My colleagues and I are not the only ones who teach these techniques. But we are the leaders in mind/body infertility research. We were the first to publish substantive research on the topic and the first to establish a world-renowned mind/body infertility program. Other such programs now exist throughout the United States and other countries, but most are modeled on our program and our research.

Our program consists of nine regular sessions of two and a half hours, plus one full-day session. Husbands attend the first, seventh, and ninth sessions. Each meeting is preceded by an optional half hour of sharing and support time that gives the approximately sixteen members of the group an informal opportunity to get to know each other and to talk among themselves. Group support is a vital part of our program, and the better the women in the group know each other, the more supportive they can be.

During the first session all the women are introduced to one another, and each one is paired off with a buddy who lives near her. Buddy pairs call each other at least once a week. Also during the first session participants learn about the relaxation response and how to elicit it. We spend a lot of time during the entire program practicing relaxation. During subsequent sessions participants learn many other coping skills.

 Mind/Body Connections

The techniques and strategies I'll show you in the following pages come from the field of mind/body medicine. Mind/body medicine is any method in which we use our minds to change our behavior or physiology in order to promote health or recover from illness. These approaches include the following:

- Any technique that induces the relaxation response, including meditation, yoga, mindfulness, deep breathing, repetitive prayer, body scan, progressive muscle relaxation, autogenic training, and guided imagery.
- Cognitive therapy, either within a group or on a one-to-one basis with a therapist. This approach allows you to challenge and replace thought patterns that trigger or reinforce depression, anxiety, and other negative emotional states.
- Coping skills that manage stress, such as self-nurturance, social support, problem solving, emotional expression, and journaling.
- Assertiveness training and communication skills that empower us to develop and sustain a nurturing network of relationships.

Relaxation and the other strategies I teach are not just tools for coping with infertility. They are life skills that can help with just about any medical or life condition. Even healthy women can benefit from learning and using these skills, because they help us to cope with the stress of anything from a flat tire to a friend's cancer diagnosis. When we can no longer cope with the stress in our lives, we feel anxious, terrified, angry, helpless, and depressed. We can't always stop the causes of our stress, but we can change our reaction to stress by using mind/body techniques. Which method you choose in order to elicit the relaxation response doesn't matter, although it's important to remember that activities such as watching TV or napping will not work, because although we think of them as being calming, they don't trigger the relaxation response.

These strategies are not a panacea. On its own, mind/body medicine is rarely a cure for disease. But used in combination with conventional medicine, mind/body medicine can help the mind to help the body. These strategies are also not quick fixes. Developing each skill requires commitment and practice. But once you learn them, you will likely find, as nearly all of my patients do, that they go a long way in helping you to cope with the pain of infertility. Not only that, but they will help you in other parts of your life as well—at work, within your marriage, and in just about any other situation that causes stress and anxiety.

Relaxation Methods

In our program we offer a variety of roads to relaxation, because not every technique is right for every person. Which is best for you? That depends on your personality, your personal history, your attention span, your stress levels. For example, some patients like directed techniques with clear-cut instructions, such as progressive muscle relaxation. Others are more drawn to techniques such as visualization, which depends on the free exercise of one's imagination. Try them all and stick with what works best for you, or jump around from one to the other based on how your focus and mood change from day to day.

I recommend that you practice some form of relaxation for about twenty minutes every day, and twice a day if you're especially stressed. Beyond that, follow your instincts. Experiment with various approaches. Weave several together, if that works. If your partner is interested, relax with him, or share this book with him and he can teach himself. Do whatever works for you.

It is entirely possible for you to teach yourself various relaxation strategies by using this book alone. Some women find it helpful, however, to practice their relaxation in a class. Relaxation classes are held in many hospitals, yoga centers, women's health centers, and community education departments.

Some people find that relaxation tapes are tremendously helpful. Relaxation requires a clearing of the mind, and when you have a lot going on in your head, it can be hard to quiet your mind thoroughly enough for breath-focus relaxation, body scan, or other relaxation exercises. They're not for everybody, but some women rave about relaxation tapes because they help listeners to shut out the world and focus more deeply on relaxation. (Our guided-relaxation audiotapes are available from the Mind/Body Medical Institute—see the Resources appendix for ordering information, or visit www.mbmi.org.)

Before you begin your relaxation session, find a quiet place in your home that can serve as your daily relaxation haven. You can decorate it beautifully with plants, paintings, statues, or just leave it as is. Be sure it's a place where you won't be disturbed by telephones, televisions, barking dogs, or other interruptions. If you have pets that won't leave you alone, shut them in another room.

Find a regular time of day to relax. Early morning is many patients' favorite time because relaxing then can set a tone of tranquillity for the day. Night can also be a good time, particularly if you suffer from insomnia. You're more likely to build relaxation into a daily habit if you do it at the same time each day, but if that's impossible, squeeze it in wherever it fits. Try to do it every day, but if you miss one day, don't judge yourself; simply try to do it the next day. Before you begin to develop a relaxation ritual, you must accept on a deep level that you deserve to take twenty minutes out of your day for mental and psychological relaxation. You need it, you have a right to it, and it will benefit your health. Don't feel guilty about telling your husband or others that you need this time for yourself—after all, it will indirectly benefit those around you.

You may choose any comfortable position for relaxation. Most patients find that sitting is best, because if you're sitting, you're less likely to fall asleep. But if you're more comfortable lying down and you won't drift off into a nap, go ahead and lie down. And you can relax with your eyes open or shut, whichever you prefer. You can even relax in a taxi, as my patient Lucy discovered:

Lucy felt very nervous about her first egg retrieval, so she decided to listen to her relaxation tape in the taxi during the thirty-minute ride to the clinic. "I said to the driver, 'Here are the directions to the clinic. I'm going to listen to this tape, and don't even think about talking to me.' So

I listened to my tape, and when it was over, we were just turning the corner into the clinic, and I calmly got out of the car. When I checked in with the nurse, I told her, 'I'm really very scared.' As she took my blood pressure, I said, 'It's probably through the roof!' and she said, 'Actually, it's normal.' I think that even though my mind was frightened, my body wasn't. I didn't have a racing heart or high blood pressure; I didn't feel like I was going to throw up or pass out. Then, in the recovery room after the retrieval, I could just close my eyes and hear the words on my tape, and that calmed me down. I did breath-focus relaxation and mini-relaxations, and I would also focus on my hands—anything to keep me from letting my fears get the best of me."

Are you ready to relax? Let me guide you through the nine kinds of relaxation techniques that I teach my infertile patients and that you can add to your toolbox of coping skills. In this chapter I will tell you how to do each of these techniques. Then, in this chapter and throughout the book, I will give you examples of how they can help you cope with and survive the various stresses of infertility.

BREATH FOCUS

When you're stressed, you tend to breathe very shallowly or to hold your breath. And as a woman—thanks to the influence of our flat-stomach-obsessed society—you probably pull in your stomach automatically throughout the day. This is an unfortunate combination of habits that leads to shallow chest breathing rather than deep abdominal breathing. As a chest breather you limit the movement of your diaphragm, the sheath of muscle that contracts during deep breathing to make ample room for the lungs to expand fully. When you engage in shallow chest breathing, the diaphragm is nearly frozen in place and your lungs don't fully expand with air as you inhale. This cuts down on the amount of oxygen delivered to your body's cells. As your body struggles to get the oxygen it needs, your heart rate and blood pressure increase. By breathing deeply and slowly—a technique called breath focus—you let go of stress and increase oxygen uptake in the cells.

Breath focus is one of the simplest relaxation techniques, and many of my patients find it very effective. People with asthma, allergies, head colds, or other breathing difficulties may find it anxiety producing, however.

How to Do Breath Focus

- Begin by taking a normal breath. Then take a deep, slow breath, allowing air to come through your nose and move deeply into your lower belly. Then breathe out through your mouth.
- Alternate normal breaths and deep breaths several times. As you do so, focus on your breathing and notice the sensations you feel with each inhalation and exhalation. Pay attention to the difference between normal breaths and deep breaths—you may begin to observe that your normal breathing is constricted and your deep breathing fosters a sense of relaxation.
- Now take a few minutes to practice deep breathing. Let the inhalations expand your belly. Allow yourself to sigh as you exhale. Repeat for several minutes.
- For the last ten minutes of breath focus, imagine that the air you breathe into your nose carries with it a sense of peace and calm and that the air you exhale is removing tension and anxiety. You may want to say to yourself, on the inhalation, "Breathing in peace and calm," and, on the exhalation, "Breathing out tension and anxiety."

BODY SCAN

When you're tense, your muscles tighten, your jaw clenches, and your face frowns. Everyone holds tension in different places—when I'm stressed, for example, I hunch up my shoulders until they're practically sticking into my ears. Body scan is a relaxation technique that allows you to determine where in your body you're holding tension and then to let go of it.

Some women are so stressed by infertility that they actually develop physical ailments such as irritable bowel syndrome (IBS) and insomnia. Body scan and other relaxation techniques can help ease or even eliminate those ailments, as my patient Becky discovered.

In addition to depression, Becky developed a terrible case of IBS, a painful ailment that forces sufferers to stay close to a bathroom. "I was terrified to go someplace new, because if I had a problem, I didn't know if I'd be able to find a bathroom. I panicked over that," Becky said. "Sometimes it was hard to leave the house. It was debilitating." Anxiety over infertility also triggered horrible tension headaches that would sometimes leave Becky bedridden. Her symptoms began to disappear, however, when she started doing daily body scan relaxation. "Listening to my body-scan tape took away my symptoms. Just by relaxing every part of my body, I felt better. By having twenty to thirty minutes a day, every single day, that were for me, with no interruptions, no obligations, relaxed me so much." Becky also found that after several weeks of doing daily body-scan relaxation, she learned to recapture that feeling of relaxation in seconds when she started to feel panicked. Both her IBS and her tension headaches have virtually disappeared.

Body scan is also helpful for people who are troubled by insomnia, as many infertile women are—it can help ease you into sleep at bedtime or after a middle-of-the-night wake-up. Patients whose minds tend to wander easily also find body scan helpful, because it gives their mind a clear purpose during relaxation.

How to Do Body Scan

- Begin with a few minutes of deep breathing. Allow your stomach to expand as you inhale and to contract as you exhale.
- Now move on to the body scan. Start by concentrating on your forehead. As you breathe in, note the way the muscles of your forehead feel. Let yourself become aware of any tension in the forehead muscles. Then, as you breathe out, let go

of muscle tension in the forehead. Continue this practice—becoming aware of tension on the in breath and letting go of tension on the out breath—for several slow, deep breaths. Remember to take nice, slow, deep breaths that cause your stomach to rise as you inhale and fall as you exhale.

- Proceed to scan the rest of your body, repeating the process of awareness of tension while inhaling, letting go of tension while exhaling. Moving from the forehead, scan the area around the eyes, then the mouth, jaw, neck, back (from the top of the spine to the tailbone), shoulders, upper arms, lower arms, hands and fingers, chest, and stomach.
- Before continuing to your lower body, do a quick mental check of your upper body. If you notice any tension between your forehead and your waist, concentrate on it as you inhale and let it go as you exhale.
- Proceed to scan your lower body—pelvis and buttocks, upper legs, lower legs, ankles, and feet. End with a scan of your entire body. If any areas of tension remain, repeat the process of concentration and letting go.

PROGRESSIVE MUSCLE RELAXATION (PMR)

Progressive muscle relaxation is a more directed version of body scan. It is similar to body scan in that you take inventory of the tension in each part of your body and then let go of that tension. With PMR, however, instead of just becoming aware of each body part, you actually increase muscle tension there before releasing it. This helps to maximize your awareness of the tension in each area of your body and intensifies the feeling of release when you let go of tension.

PMR is especially effective for people with very active minds, who have trouble focusing and tend to prefer a more orchestrated form of relaxation.

How to Do Progressive Muscle Relaxation

- Prepare yourself for PMR by breathing deeply for a few minutes.
- Start PMR by concentrating on your forehead. Consciously tighten the muscles of your forehead while counting slowly from one to five. Hold your forehead muscles as tight as you can for the duration of this count. Then let go of your tense forehead muscles while taking a nice, slow, deep breath. Notice your stomach rise as you inhale and fall back down as you exhale.
- Repeat with the forehead muscles.
- Now move down to your eyes and repeat the process twice.
- Continue throughout the body, tightening the muscles in a particular area for a count of one to five, releasing the tension as you take a slow, deep, breath and repeating the process. Move from your eyes to your jaw, neck, back (from the top of the spine to the tailbone), right shoulder (bring it up as high as you can), upper right arm, right forearm, right hand (tighten it into a fist), left shoulder, upper left arm, left forearm, left hand. Move to the chest, abdomen, pelvis and buttocks, upper right leg, lower right leg, right foot (point your toes up), upper left leg, lower left leg, left foot.
- As your relaxation comes to a close, do a mental check of your entire body, head to toes. If you notice any remaining areas of tension, tense those muscles for a count of five and then let go of the tension in those muscles as you take slow, deep breaths.

MEDITATION

Meditation is a spiritual practice dating back thousands of years. My mentor Herbert Benson, M.D., has found that meditation can elicit the relaxation response when it includes these two basic elements: First, you turn your attention inward and focus repetitively on your breathing and a simple word, phrase, or prayer. And second, you adopt a nonjudgmental attitude toward any thoughts or feelings that float through your consciousness.

When practiced regularly, meditation can clear the mind of emotional clutter, increase your sense of inner peace, and foster spiritual connectedness, particularly if your focus word or phrase has religious meaning.

Meditation works well for people who are able to quiet their minds. If you have a hyperactive, racing mind, you may not be able to still yourself enough to meditate. Don't assume, though, that you can't meditate just because you've got a lot going on in your head. I've had patients who loved meditation because it was the only way they had ever found to slow down. Betsy, a patient with secondary infertility, is a good example.

> *My biggest problem was insomnia. It was dreadful—just the worst thing. I would be up all the time, and I was a wreck. All day I would dread the evening, and at night I'd think, "Here we go again, I have to battle this demon again." I just couldn't shut off my brain. There's such a chatter going on in my head all the time. But meditating made a huge difference for me. It would calm me down, and I could sleep. Meditating saved my brain and my life, as far as I'm concerned.*

How to Meditate

- Choose a word or phrase for your meditation. Pick something that has meaning to you, such as "peace and calm" or "serenity." For this instruction we'll use the words "peace and calm."
- Close your eyes, if you're comfortable doing so. If not, keep them open.
- Starting with the number ten, count down to zero, one number for each breath you take. Notice that your breathing may get slower as you count down.
- As you breathe in, begin to concentrate on the word "peace" in your mind. As you exhale, concentrate on the word

"calm." Inhale through your nose and exhale through your mouth, if that feels comfortable.

- If your attention wanders, gently bring it back to the words on which you're focusing. If thoughts or feelings intrude on your practice, acknowledge them gently—don't encourage them or push them away—and then return to your breathing.
- Gradually slow your breathing by pausing a few seconds after you inhale and again after exhaling.
- As your time for meditation comes to an end, continue to be aware of your breathing, but start to be aware of where you are, the sounds around you, and where you are sitting. When you feel ready, open your eyes, look down for a few minutes, and get up slowly.

PRAYERFUL MEDITATION

Numerous scientific studies have shown that prayer and affiliation with a spiritual community can aid in physical healing. Whether prayer helps by reducing anxiety, connecting us to others, or invoking the healing energy of a higher power, we don't know. But we do know that it can help some people feel better, and prayerful meditation can be an effective way to elicit the relaxation response.

Prayerful meditation is most effective for people who are comforted by spirituality and who believe in God or another higher power. Those without a deep religious commitment may find prayer antagonizing, or, to their surprise, they may discover that it is beneficial and renews within them a desire to return to a religious practice that they had left behind.

How to Do Prayerful Meditation

- Choose a focus word or phrase that has a personal religious or spiritual meaning to you, such as "Come, Lord" or "Our Father" or "*Sh'ma Yisroel*" ("Hear, O Israel") or "*Shantih*" ("Peace") or "Allah."
- Proceed as for meditation.

MINDFULNESS

So often we live in the past or the future. We dwell on distant hurts or worries, or we fret about what is to come later in the day, next week, next year. When we do this, the here and now slips away, and we live in a state of mindlessness.

A practice called mindfulness can help pull you back into the moment. Being mindful means appreciating the here and now—the soft breeze on your face, the smell of stew cooking in the kitchen, the happy feeling you experience when your husband surprises you with flowers.

Mindfulness is important for infertile women for several reasons. Because of the cyclical nature of reproduction, you are constantly thinking of past and future—the period you had last month, the period you're hoping not to get this month, the medical procedure scheduled for next week. What's more, infertility is so stressful that it can overshadow everything else in your life—even the good things. Mindfulness can help you remember that even though you've got a terrible, horrible stressor in your life, it is not the *only* thing. That's how it worked for Karen, who participated in my program after three unsuccessful IUIs and two failed IVFs.

> *Mindfulness taught me to try to refocus on all of the wonderful things I do have rather than what I don't have. It's so easy to look at what everyone else has and say, "Oh, they're so lucky, they have three kids," or whatever. When I take a step back and look at what I have, I can say, "Well, I'm pretty lucky, too. I have this, this, and this that they don't have." I have a great family, wonderful friends, and a job I love. It's so important for me to remember those things instead of getting caught up in the whole pregnancy thing all the time. Now I'm more laid back about it.*

You can practice mindfulness while you're engaged in an activity or as a form of relaxation. Although mindful meditation isn't for everyone—those with a limited ability to empty their minds don't always do well with mindful meditation—anyone can benefit from taking steps to be fully present in her own life.

How to Practice Mindful Activity

You can be mindful while doing any activity, from walking through a beautiful state park to folding laundry. The trick is to be fully engaged in your activity, training all of your senses on what you're doing.

Here's a great way to practice mindfulness—and an assignment I'm sure you won't mind doing. Try eating a spoonful of ice cream mindfully. Slowly open the freezer, feel the puff of cold air on your face, take the ice cream carton out of the freezer, and set it on the counter. Open the carton, scoop out a spoonful, and eat it, paying close attention to every sensory detail—the sound of the lid coming off the container, the ice cream's coldness on your tongue, its flavor and sweetness and smoothness in your mouth, its change in texture as it softens and melts. Try to extract as much sensory pleasure as possible from the experience.

Practice that same mindfulness with ordinary activities throughout the day. When you make yourself a cup of tea, pay close attention to the smell of the tea bag, the sound of the water pouring into the cup, the warmth of the cup in your hands, the deep brownish-orange color of the brew. Notice the smell of your husband's aftershave, the crackle of your breakfast cereal, the feel of the woolly mittens you pull onto your hands, the rumble of the subway you ride to work. My patient Kerry focused on a tree.

> *I was very mindful this spring of going to bed with my window shades up so that the first thing I would see when I woke up was my dogwood tree in bloom. For two weeks, every morning was fantastic because I would roll over and there was my beautiful tree, and it was all I could see.*

Mindfulness extends beyond what you hear, taste, and smell, however. Being mindful of the small pleasures in life can extend to people, places, and situations. Try being aware of what makes you happy in your life—your relationship with your brother, volunteer work that fulfills you, your sunny backyard. Appreciating these gifts can expand your scope and put infertility into a less hurtful perspective.

Among the most useful mindful activities are mindful walks. These can be a wonderful way to secure yourself in the present moment, and they are a particularly helpful activity for infertile women. They are pleasurable, relaxing, emotionally energizing, and an effective way to take your mind away from worries and anxieties.

How to Take a Mindful Walk

- Choose the most pleasant route you can think of—a pathway by a reservoir, a trail through a park, a boardwalk along the ocean.
- Walk slowly, fully experiencing the sensations of walking one step at a time.
- Focus on allowing each of your senses to take in the surroundings. Smell the aroma of grass, ocean mist, or whatever scents waft by. Notice the sights of trees, flowers, people. Really listen to the sounds of birds chirping, branches rustling, your feet crunching on crispy fall leaves. Notice the feeling of your feet landing rhythmically on the pavement or gravel.
- If negative thoughts intrude on your awareness, gently return to your focus on the sensations of walking and the sights, sounds, and smells around you. When these thoughts intrude, don't judge yourself, just nudge them away and return to your sensory exploration of the world around you.
- Walk for as long as you like, but don't think of this as an exercise walk. Leave your Walkman and your heart-rate monitor at home, and focus only on the sensations around you.

How to Practice Mindful Meditation to Elicit the Relaxation Response

- Begin by focusing on your breathing. You may wish to use a focus word or phrase, but it is not necessary. If you don't, just focus on the sensations of your breathing—your belly as it rises and falls, the air as it enters your nostrils and leaves your mouth.
- If you notice that thoughts, worries, hopes, fears, and fantasies keep appearing, don't worry. That is a natural process.

As you sit in stillness, your body in a state of quiet and relaxation, watch each thought as it comes and goes. Be mindful of the process of thinking. Notice how the thoughts are always subtly shifting, moving, dissolving.

- When you notice that you've been carried away in a stream of thoughts, observe that this has happened. Without judging yourself, gently turn your awareness to your breath, allowing your breath consciousness to be in the foreground and your thoughts in the background. The breath is the most natural way to center yourself and be anchored in the present moment.
- For the remaining time, keep your breathing in the foreground of your consciousness. To the best of your ability, keep whatever else may arise—sensations in the body, thoughts in the mind, sounds in the environment—in the background. If they do intrude, don't struggle with them. Rather, be aware of them and simply return to your breathing.
- As you complete your mindfulness meditation, slowly bring your awareness back to your surroundings.

One of the goals of these exercises is to become comfortable enough with mindfulness that you can easily bring it into other parts of your day—time spent at work, with friends and family, even driving in the car. Mindfulness allows you to experience the joy of small moments even when you are continually stressed by larger events. Rediscovering the happiness of everyday activities can help remind you that although you have one enormous stressor in your life—infertility—there are still many smaller things that can bring you pleasure.

GUIDED IMAGERY

Guided imagery uses mental pictures of scenes, places, and experiences that evoke a sense of inner calm and peace. You can create these mental pictures for yourself or listen to audiotapes that take you to a tranquil place, such as a field of flowers or a sunny beach.

Guided imagery is an excellent relaxation technique for anyone with a strong imagination who enjoys a more freeform type of relaxation.

How to Do Guided Imagery

- Choose a visualization location. Remember that it's important to pick wisely and to avoid placing yourself in a destination that might upset you. For example, a woman with hay fever may become anxious visualizing herself in a field of flowers. Some infertile women find comfort imagining themselves being pregnant or holding a baby in their arms; however, others find this image upsetting. Experiment with different destinations and use what works best for you.
- Take several slow, deep, cleansing breaths.
- Go in your mind to a special place that you love, a place where you have felt relaxed or know that you would feel relaxed. It can be a favorite vacation spot, your own backyard, a place you've seen in the movies or read about in a book.
- Spend time in this place, envisioning yourself sitting, standing, moving, or lying—whatever feels right. Take in the sensations all around you.
- Focus on smells, sounds, movement of clouds, shapes and colors, the feel of the air on your skin or the grass on your feet, and so on. Allow yourself to become completely absorbed in the sensual aspects of these images.
- If your concentration is interrupted by anxious or disturbing thoughts or images, acknowledge them and then return gently to the specific sights, sounds, and smells that surround you.

Guided imagery and other relaxation techniques allowed my patient Helen to slow down for the first time in years.

My mind was constantly running full steam ahead, six steps beyond where I am. I'd be in the shower physically, but mentally I was already in the car going to work. I was always ahead of myself. My cruising speed was very high—my nose was forever a millionth of an inch away from the wall, so it didn't take much for me to smash into it. Relaxation and mindfulness helped to slow down my whole engine. I listened to my relaxation tapes a couple times a day, and it really helped me. Then sometimes, during the day, I would just stop and think about what I was doing—like while I was taking a shower, I would stop and pay attention to the smell of the shampoo or the feel of the warm water. In the car I would stop and really listen to the music that was playing, as opposed to thinking about the traffic or what I was going to do at work that day. Now my tendency to go ninety miles per hour returns sometimes, but I recognize it and try to slow down. I've done my relaxation tapes so many times that I can just sit down and get that feeling of relaxation anywhere, even when I'm very stressed.

AUTOGENIC TRAINING

Autogenic training uses verbal suggestions to help you achieve a state of relaxation. The verbal suggestions, called orientations, are a subtle form of self-hypnosis. When you use autogenic training, you are essentially bypassing your conscious mind in order to guide your body to relax.

Autogenic training can be very helpful if you have trouble doing breath focus or meditation. It gives your mind a very clear focus, making it less susceptible to intrusive thoughts. Women with infertility find this an effective form of relaxation, because their condition subjects them to such intense stress that it often interferes with the concentration needed for relaxation. Autogenic training is also useful for insomnia, because the sensation of warmth and heaviness can create a soporific, sleep-inducing feeling.

How to Do Autogenic Training

- Autogenic training requires a helper (or a tape recorder to record your own voice), so start by lining up an assistant. Have your partner read (or make a recording of) these instructions, adapted from Herbert Benson, M.D., "The Relaxation

Response," in *Mind-Body Medicine*, edited by Daniel Goleman and Joel Gurin (Consumer Reports Books, 1993).

- Focus on the sensations of breathing. Imagine your breath rolling in and out like ocean waves. Think, "My breath is calm and effortless . . . calm and effortless. . . ." Repeat the phrase to yourself as you imagine waves of relaxation flowing through your body: through your chest and shoulders, into your arms and back, into your hips and legs. Feel a sense of tranquillity moving through your entire body. Continue for several minutes.
- Now focus on your arms and hands. Think, "My arms are heavy and warm. Warmth is flowing through my arms, into my wrists, hands, and fingers. My arms and hands are heavy and warm." Stay with these thoughts and the feelings in your arms and hands for several minutes.
- Now bring your focus to your legs. Imagine warmth and heaviness flowing from your arms down into your legs. Think, "My legs are becoming heavy and warm. Warmth is flowing through my feet . . . down into my toes. My legs and feet are heavy and warm." Stay with these thoughts and feelings in your legs and feet for a few minutes.
- Now scan your body for any points of tension, and if you find some, let them go limp, your muscles relaxed. Notice how heavy, warm, and limp your body has become. Think, "All my muscles are letting go. I'm getting more and more relaxed."
- Finally, take a deep breath, noticing how the air fills your lungs and moves down into your abdomen. As you breathe out, think, "I am calm. . . . I am calm. . . ." Do this for a few minutes, feeling the peacefulness throughout your body.
- As your practice session ends, count to three, taking a deep breath and exhaling with each number. Open your eyes and get up slowly. Stretch before returning to your everyday activities.

YOGA

Yoga is an ancient practice that combines deep breathing, meditation, and physical postures. Although there are many different types of yoga, all share a common goal of bringing peace and tranquillity to the body.

Many of my infertility patients have fallen in love with yoga. Not only does it calm their troubled minds, but it makes their bodies feel great, too. Infertile women are often told to limit their exercise, which makes them feel out of shape and out of control. What's more, medical treatments, injections, and diagnostic tests can leave them feeling as if their body belongs not to them but to their doctors! Yoga gives infertile women a way to exercise safely and to regain some control of their body. It can also be very helpful to women who are angry with their body for not being able to conceive, because yoga gives them a way to reconnect with and honor their antagonized body. Yoga can help you experience yourself as a whole person.

Yoga is especially good if your mind is so busy and full that you have trouble slowing down long enough to elicit the relaxation response in other ways. By releasing muscle tension, yoga can quiet the mind in a way that other relaxation techniques cannot.

Anyone can practice yoga, even if you're out of shape or injured. Indeed, people who sign up for yoga classes are often nursing injuries of some kind. Although there are many books available that teach yoga postures, you're better off taking a class from a knowledgeable instructor once a week or so and practicing the postures at home. Or if you can't squeeze classes into your schedule, try a yoga video. For an excellent resource on yoga poses, pick up the book *Yoga for Dummies* by Georg Feuerstein, Ph.D., and Larry Payne, Ph.D. (Hungry Minds, 1999).

Many health clubs, YWCAs, spas, community centers, hospitals, and HMOs offer yoga classes. The various styles of yoga teach similar positions, but they differ in their intensity, spanning a range from relaxing to athletic; some incorporate strength and cardiovascular training as well as stretching and meditation. I usually recommend that infertility patients choose less-athletic yoga practices; however, your doctor should make the final call on what's best for you.

Generally, hatha yoga requires only mild exertion; that is what we teach our infertility patients. One hatha class can differ greatly from the next, with one being more relaxed and another being more challenging. There are other types of yoga, as well; again, although actual practice varies from instructor to instructor, I think some yoga styles are better for infertile women than others.

- Iyengar is the most precise approach. Iyengar students focus heavily on perfecting the asanas, or positions, paying close attention to the structure of the positions, rather than concentrating on vigor, although Iyengar yoga can also be practiced athletically.
- Kripalu tends to focus on the internal, subjective, emotional aspects of the body. Perfecting the postures is less important than is connecting with your inner world.
- Astanga (also known as power yoga) is a sophisticated series of interconnected yoga postures laced together with transitional movements. Astanga yoga, which can be compared to a gymnastic workout, generates great heat within the body that helps warm the muscles and increase flexibility. An astanga class is likely to be more like an exercise class than a time for relaxation and meditation. If your doctor has recommended that you stick to less-vigorous exercise, astanga yoga may be too vigorous for you.
- Bikram (also known as hot yoga) is a series of linked postures done in a room heated to 105 degrees. The heat of the room helps the connective tissues to open up and fosters greater gains in flexibility, strength, and healing. Some people find the heat intolerable, however, and I don't recommend this for infertility patients.
- Kundalini includes some postures but focuses more on breathing, energy gathering, spiritual awakening, and emotional healing.

When you're picking a yoga class, look for one with an experienced instructor and a small class—preferably twelve to fifteen people. The instructor should make you feel safe, should be sensitive and willing to listen to you, and shouldn't be pushy or militant. A good instructor can make or break a yoga class. Teachers interpret their own schools of yoga in different ways—for example, you may love one instructor's take on kundalini and dislike the next one's. The best way to find a class that's right for you is to give it a test drive; if you don't like it, try another.

If your partner is up for it, the two of you can try doing yoga together as a couple. Couples yoga allows you to do something physical together that isn't sexual—it allows you to comfort and nurture each other in a nonsexual way. Sex often provokes anxiety for infertile couples. Yoga reduces anxiety and gives husband and wife a low-pressure way to do something sensual together. I'll talk about couples yoga in greater detail in Chapter 4, but for now let's look at the experience of Sandra, a thirty-four-year-old health-care manager who has been trying to get pregnant for three years. Because she suffered from a rare condition in which her body stopped menstruating and ovulating, Sandra's doctors suggested that she reduce the intensity of her workouts and limit herself to walking and yoga. This radically changed her relationship with her husband.

> *I used to exercise every day. It was a really important part of my life. Tom and I played tennis together, we biked together, we went running together, we hiked together. And I can't do that anymore. Suddenly things that we were doing together, he's doing with other people. And that sucks. Now I'm walking with girlfriends, which is fun, but it's not what I'd prefer to do. My husband and I are losing quality time together because we're not doing those things.*

During the Sunday session of the infertility program, we invite husbands to join their wives in trying couples yoga. Sometimes the men don't like it—they're embarrassed to get down on the floor and move their bodies, or they aren't flexible enough to do the moves. But Sandra's husband, Tom, is an athlete, and he warmed right up to the couples yoga. It turned out to be a perfect way for Sandra and Tom to regain some of the closeness they'd lost when she had to reign in her physical activity.

He really enjoys it, and so do I. Yoga is new to both of us, so it's something we're learning together. We now go three times a week to a yoga studio that's close to our home.

 Other Coping Skills

Daily elicitation of the relaxation response is a wonderful way to reduce stress, but it is certainly not the only coping skill in the toolbox. Just as various house repairs call for different tools, the daily emotional challenges of infertility require a wide range of coping skills. In this section I'll describe and explain several other techniques that will help you survive the stresses of infertility.

MINI-RELAXATIONS

We talked before about how stress and anxiety cause us to breathe shallowly. When something upsetting happens—say, when a friend calls to announce that she's pregnant—you immediately begin to restrict your breathing and automatically engage in shallow chest breathing rather than deep abdominal breathing. Unless you consciously change your breathing pattern, a respiratory chain of events begins to unfold. Less oxygen goes into your lungs, which sound an alarm in your brain and body. The fight-or-flight response kicks in, even if stress is stable or recedes. A vicious cycle ensues: Fight-or-flight and its attendant stress hormones keep you anxious and sustain the shallow breathing. Your diaphragm remains stuck, and so does your state of agitation. You can't think straight, and you experience a variety of emotional and physical symptoms. You lose oxygen, energy, and perspective. Over time, shallow breathing results in emotional distress and physical exhaustion.

You can stop that chain of events immediately with a mini-relaxation. Mini-relaxations, or "minis" for short, are among the most useful coping tools. In just a few seconds a mini can shift you from shallow chest breathing to deep abdominal breathing, speeding oxygen to your cells and giving you a chance to step back, gather your emotional resources, and cope with whatever challenge faces you. Minis instantaneously help you break the vicious cycle of tension and anxiety. They did for Dana, who after years of infertility is planning to adopt.

> *One of my stress things is that I stop breathing. When I was tense, I would hold my breath. It was especially bad when I was driving on the highway to my doctor's office—my anxiety level was through the roof. But now I know how to stop and look at myself throughout the day and to check my breathing by doing a mini. Just learning to recognize when I was stressed, being able to be aware that I was holding my breath, was so valuable.*

Most of my infertile patients love minis. Some use minis dozens of times a day to give them a psychological and physical lift anywhere, anytime, and in any circumstance. Within minutes, minis can take the edge off their anxiety, quiet their racing physiology, clear their minds, and revive their sense of control. Minis are little reminders throughout the day that you can find your center, even when events or other people seem completely out of control.

You can do a mini anytime you don't have time for a full relaxation session. You can do one with your eyes opened or closed, in the presence of others, even while you're driving. Here are some great times to do minis:

- Before and during blood tests
- Before and during injections
- During any stressful phase of preparation for medical tests or procedures
- Before and during ultrasounds
- Before calling the doctor's office for test results
- When you're waiting for your doctor to call you back
- When you're put on hold by your doctor's office
- When you see a pregnant woman in the grocery store
- When an invitation to a baby shower arrives in the mail
- When your mother-in-law says, "If you'd just stop trying, you'd get pregnant!"

- When your neighbor says, "When are you going to have a baby? You're not getting any younger, you know!"
- Anytime you go to the bathroom toward the end of your cycle
- If you've suffered miscarriages and are pregnant again, anytime you go to the bathroom
- When you walk past a baby-clothes shop in the mall
- When your pregnant sister-in-law e-mails you her ultrasound photos

Here's what Cara has to say about them:

> *Minis save me during medical tests. I do them when I have to sit and wait to get my blood drawn, or when I get the phlebotomist who isn't very good and leaves bruises—you get to know which ones are the good ones and which are the bad ones. Minis allow me to step back and take a breather, assess the situation, bring my blood pressure back down, and then move on.*

And Ginny:

> *I use minis in the middle of the night. I often wake up in a panic between three and five* A.M.*, obsessing about medical details—did they put me on enough medication? Am I going to make enough follicles? Minis help me to stop obsessing and go back to sleep.*

Don't wait until after a stressful situation occurs to do a mini—you can also use minis to prevent emotional distress by doing one before an event that you know is likely to make you tense, anxious, or upset. For example, doing a mini before a doctor's appointment can help calm your fears before you even walk into the office.

I teach four different versions of mini-relaxations. Each can be an incredibly helpful way of coping on the fly.

How to Do Mini Version 1

- Sit down or, preferably, lie down in a comfortable position. Take a deep, slow breath. Notice any movement in your chest and abdomen. Place a hand on your abdomen, just on top of your belly button. Allow your abdomen to rise about an inch as you inhale. As you exhale, notice that your abdomen will fall about an inch. Also notice that your chest will rise slightly at the same time that your abdomen rises.
- Become aware of your diaphragm, moving down as you inhale, back up as you exhale. Remember that it is impossible to breathe abdominally if your diaphragm does not move. And it is impossible to let your diaphragm move if your stomach muscles are tight. So relax your stomach muscles! If you are having trouble, try breathing in through your nose and out through your mouth. Enjoy the sensations of abdominal breathing for several breaths, for as long as you desire.

Mini Version 2

- Count down from ten to zero while taking one complete breath—one inhalation, one exhalation—with each number.
- If you start to feel light-headed or dizzy, slow down your counting. When you get to zero, you should feel better. If not, try doing it again.

Mini Version 3

- As you inhale, count very slowly from one to four.
- As you exhale, count slowly back down, from four to one.
- Do this for several breaths.

Mini Version 4

- Use any of the other three methods as you breathe: Simply breathe as you feel your stomach rise, add a ten-to-zero count with each breath, or add a one-to-four/four-to-one count as you inhale and exhale.
- But this time, regardless of what method you use, pause for a few seconds after each in breath. Pause again for a few seconds after each out breath.
- Do this for several breaths, or for as long as you wish.

COGNITIVE RESTRUCTURING

Cognitive restructuring is based on a very simple but true principle: Our thoughts can determine our emotional states, and our emotional states can influence our physical health.

Let me give you an example of this: If you constantly think that if you can't have a biological child you will be a worthless failure, you are likely to become seriously depressed and anxious. If these thoughts continue unchecked, and your depression and anxiety become chronic, you will also become vulnerable to a host of physical ailments.

Cognitive restructuring interrupts this cycle of emotional thoughts that negatively affect your physical health. When you use cognitive restructuring, you identify your negative thoughts, question their veracity and validity, and then replace them with new thoughts that are kinder and more accurate.

All of us have certain recurring thoughts or phrases that we say to ourselves repeatedly and unconsciously, almost like a tape loop that plays over and over in our heads. "I'm too fat" and "I'm not good enough" and "If I'm not perfect, I'm no good at all" are among the most universal. For women struggling with infertility, the tape loops may include "I'm never going to have a baby!" or "I can't get pregnant because God thinks I would be a bad mother," or "It's all my fault that we can't get pregnant."

Cognitive restructuring helps you to free yourself from the pain that destructive thoughts bring you by putting negative thought loops to the test using four basic questions. Then you can erase those negative tapes and record new, more constructive ones.

The method of cognitive restructuring I teach is quite simple, although the process itself isn't always easy. It begins with those four basic questions, which I have adapted from the work of leading practitioners in cognitive therapy.

First, identify one common negative thought pattern, one tape loop that plays repeatedly in your head. Now ask yourself the following questions about that negative thought:

1. Does this thought contribute to my stress?
2. Where did I learn this thought?
3. Is this thought logical?
4. Is this thought true?

Before you can restructure an automatic negative thought, you must first honestly confront that thought, discover its origins and effects, and put it to the test of logic.

Let's ask our four cognitive restructuring questions about this statement: "If I can't have a biological child, I am a failure."

Question 1: Does this thought contribute to my stress?
Answer: You bet it does. It's an all-or-nothing thought that leaves you feeling that the success and happiness of your entire life depend on one single thing: having a biological child.

Question 2: Where did I learn this thought?
Answer: Did you grow up with a mother who believed that her only value came from producing and raising children? Or perhaps you feel that your husband or family doesn't respect your career, your creativity, and your personality but only judges you by whether or not you have children? Search deeply for the source of this thought so that you can face it and defuse its power.

Question 3: Is this thought logical?
Answer: No, it's not. All people have worth, whether they produce children or not. There are many people who have made tremendous contributions to the world that have nothing to do with having biological children. And there are quite a few biological parents out there who contribute very little to society. Our worth is determined by our actions, not by whether we produce genetic offspring.

Question 4: Is this thought true?
Answer: No. First, you're going to do everything in your power to conceive a biological child. Then, if you can't conceive, you'll consider other pathways to parenthood. You can be a successful and happy person whether you give birth to a child, adopt a child, or choose to remain childless.

By identifying and restructuring your negative thought, you can deflate its ability to hurt you. Over time, you can teach yourself to replace the thought "I'll be a worthless failure if I don't have a biological child" with "I'm going to do my best to conceive, but if I can't, I'll find other ways to make my dreams come true and to be a success." This process helped my patient Alex, who suffered from a rare gynecological disorder that has caused her to miscarry several times.

> *When I think, "I'm never going to have kids," I restructure my thoughts and say, "Well, there is still a high probability that I'll carry a baby to term. It's going to be a longer road than I ever imagined, but I have a wonderful, loving, strong husband by my side, a great network of friends and supportive family, and a fantastic house and a happy life. I have to focus on that."*

Cognitive restructuring can also help with insomnia, which is common among infertile women and which plagued my patient Betsy, whom I mentioned earlier in this chapter. Meditation helped Betsy to calm down at bedtime, and cognitive restructuring helped her to change her thoughts about sleep.

Before I even got into bed, I would be saying to myself, "I'm not going to be able to sleep, I'm going to be up all night, I'm going to be exhausted tomorrow, I'll never sleep again." I got so far down that path before I even put my head on the pillow! The process of cognitive restructuring allowed me to see that just thinking those things was probably contributing to my insomnia. Restructuring my worries about insomnia really made me dread bedtime less.

JOURNALING

As a teenager, you may have written in a diary, pouring out your thoughts, worries, doubts, and dreams, using writing as a way to negotiate the ups and downs of adolescence. But journal writing isn't just kid stuff. Research shows that writing about stressful events can be an extremely helpful coping strategy for anyone who's experienced a trauma. Journaling reduces stress and anxiety, and it can help you uncover truths, strengths, and solutions that may be buried within you. It is also an excellent way to air and explore volatile emotions without hurting others.

Infertility provokes a range of emotions. One of the hardest to deal with is anger, because although it is a completely justified emotion, it has the potential to cause terrible damage. For example, a woman might become angry about all the medical treatments she must undergo. This is particularly common in couples with male-factor infertility—women feel furious that they are reproductively healthy and yet their bodies are put through extremely invasive procedures to compensate for the male factor. They don't want to tell their husbands how angry they are—they love their husbands and don't want to hurt them or devastate the marriage. That's where journaling comes in. When you write about your anger, you can release it. And since you don't show your journal to anyone, you are expressing your anger in a way that won't hurt anyone.

In our mind/body infertility groups, we invite women to write for twenty minutes about the most traumatic event of their infertility. Then we recommend that they write again for at least twenty minutes on each of the next three days. (If you're on a roll, though, don't stop just because twenty minutes are up.) Be completely honest—you're writing for yourself, not for anyone else. You might want to experiment with different forms of writing. Don't limit yourself to writing long narratives in a notebook. Some people find more satisfaction in writing letters, a dialogue with another person, or a speech.

When I asked Terry and the other members of the mind/body program to write in journals about their infertility, Terry wrote about her miscarriage, and that exercise released a huge amount of emotion—and healing power. Journaling also gave her the impetus to talk with her husband about the miscarriage and make peace with it.

> *When I had my miscarriage two years ago, my husband wanted to be in the room for the D&C, but I didn't allow him to do that. I wanted to protect him and everyone else. I realized through journaling that I'd never dealt with my miscarriage and I'd never let anyone around me grieve, including my husband. After I wrote about it, I was able to uncover a lot of that emotion and start dealing with it. It will always be a part of me, but now I feel like I don't recall it as much. Before, it was always under the surface, and certain things would bring it up. But now it's not as close to the surface. Writing about it was extremely healing.*

Another form of journaling is letter writing. A number of my patients have found tremendous peace after writing letters to people who have hurt them—family members, friends, spouses, even doctors. Some send the letters, some don't. Either way, writing a letter can be an excellent outlet for letting go of feelings, as it was for my patient Faith. After her IVF failed, Faith wrote a letter to her doctor. In it she told him that she was angry with him for treating her insensitively and forcing her to wait longer than she should have for test results. "After I wrote it, I cried all day, off and on. It was so cathartic for me to write out all my feelings and then to cry like that." She wasn't sure whether to send the letter, so she waited a few days—something I highly recommend—and then sent it when she cooled off and reread it. He never responded, but that didn't matter to Faith. Just writing the letter made her feel so much better.

SELF-NURTURANCE: PRACTICING PLEASURE

When was the last time you treated yourself well—really well? If you're like most women, you probably can't even remember when you did something especially nice for yourself.

As women, we're conditioned to take care of others, to focus our energies on the needs of spouses, family members, even strangers. Because we're brought up to be selfless, we're too guilt ridden to take much time for ourselves, for our pleasures, our growth, and our development. Although we readily take care of others, we feel guilty taking care of ourselves!

Failing to nurture ourselves as we nurture others is a self-esteem issue, but it's also a health issue. Mind/body researchers have associated self-denial of various forms with autoimmune diseases and even the progression of certain types of cancer. And in my clinical practice I have seen evidence that women with low self-esteem, who always put others before themselves, are more prone to a variety of physical symptoms and disease. When these women learn to nurture themselves, they feel better, and their self-esteem and health improve.

Infertile women often feel as if their bodies belong to their doctors—they poke you and prod you and take your blood and harvest your eggs and tell you when you can and can't be intimate with your husband. By nurturing yourself, you reclaim your power to make yourself happy and to be in control of your body.

Here's my self-nurture prescription: I recommend that you take time each day to engage in some activity that is soul nourishing. This is a time to care for your own needs, to enjoy a facet of yourself, to experience pleasure via your senses.

Start by giving yourself permission to take that time. This may be the hardest step, particularly if you were brought up believing that time spent on your own pleasure was wasted time, or if you were raised by a mother who never took time for herself. If you have trouble chasing away guilt, look at the other people in your family—they probably have no qualms about taking time for themselves. Your husband may watch sports for hours on Sunday afternoons. Your father might golf every Friday morning. Let their example motivate you to take time for your own pleasure, too.

During my infertility program, I give my patients an assignment: to find ways to nurture themselves for a total of about thirty minutes a day. Some of the choices they've made include the following:

- Have a manicure or pedicure.
- Sit on a chaise longue in the backyard with a cool drink and the latest issue of your favorite magazine.
- Take an afternoon nap.
- Read a trashy novel.
- Listen to Barenaked Ladies or U2 or whatever group or singer you like, turned up *loud*.
- Plant an herb garden.
- Make some popcorn and watch a mushy movie.
- Grab a mystery novel and camp out at Starbucks for the morning (stick to decaf coffee or tea).
- Borrow a friend's puppy and romp around with it in the grass.
- Visit an art museum.
- Eat an ice cream cone (without punishing yourself with warnings about fat and calories).
- Knit, crochet, quilt, or do some other handicraft.
- Eat a wonderful piece of chocolate, guilt free.
- Go shopping—but not for clothes, because trying on clothes can remind you of weight you've gained or the fact that you're not buying maternity clothes. Buy yourself a funky hat, wild socks, or zany underwear instead.

What you choose to do doesn't matter, as long as you're indulging in a pleasure that you would normally deny yourself. Pick something that emphasizes enjoyment, play, and a sense of fun—self-nurturance should never feel like just another chore on the daily to-do list. It should also not be a financial strain—a seventy-five-dollar restaurant meal is pleasurable, certainly, but it loses its nurturing effect if you worry about having overspent. Your pleasurable activity should be something you look forward to, you revel in, and you remember fondly for the remainder of the day. Hattie chose to pamper her body.

> *I never had a pedicure in my life. And I never had a massage in my life. But I started having them, just as a way of taking care of myself and focusing some energy on myself. I'm a nurse, and I realized I was always taking care of everyone else. It felt so nice to take care of me for a change. It's not the massage or the pedicure that help me so much as the recognition that I need to nurture myself. Although those things sure feel good.*

Betty's pleasure came from being creative.

> *I started taking a painting class. It was such a great creative outlet for me. I had never painted before, but I realized I really liked it, and I was a lot better at it than I thought I would be. If I tried to nurture myself by taking a hot bath, when I closed my eyes, I would think about everything that was going on. But when I painted, I would focus on the painting and talking with the other people in the class and the teacher, and I would forget about infertility.*

On a hot summer day one of my patients bought herself a kiddie pool and filled it with cool water. Then she spent the afternoon lying in it, watching the clouds move in the sky, listening to the breeze ruffle the leaves. She said it was one of the best afternoons of her life, because she just relaxed.

Performing a self-nurturing act can often help heal the pain of a difficult event, such as a disappointing test result or the arrival of your period. Doing so can remind you that even though you aren't pregnant, you can still enjoy and savor some of life's smaller pleasures.

Sometimes nurturing yourself means giving yourself permission to say no to situations that will upset you and cause you grief—a baby shower, a Passover celebration at a fertile sister's house, your niece's kindergarten dance recital.

GROUP SUPPORT

Why is group support so valuable for infertile women? Here's what my patient Gilda, who has been trying to get pregnant for three and a half years, thinks about this:

> *I had never met anybody who had battled infertility—all of my close friends got pregnant easily. When I walked into Ali's program and saw a roomful of women who were in my age range and who have been through the same things I have, I was overwhelmed. It was truly the best thing I've ever gone through in my life. It gave me an opportunity to meet an incredible group of women and gain more knowledge, not only from the infertility standpoint but also knowledge of how to deal with what I'm going through and how to put things back in perspective. I didn't realize until I went to that class how out of control I really was. Those women helped me get back in control. Plus, I now have a whole group of people I can call if I have questions, because they've been through it all.*

Another of my patients, Laurel, feels this way about it:

> *A lot of people say things to the group that they can't tell their closest friends. We know things about each other that nobody else in the world knows. We complain about our husbands, we cry together, and we even laugh, too. We joke about the shots, we make fun of the doctors—I probably shouldn't say that, but a lot of us have the same doctors, and we joke about them. But it's also a network, an exchange of medical technology. Knowledge is power in this situation. We have to do something to take control.*

Research shows that group support can help people with just about any health problem to feel better. In several landmark studies, women with cancer who participated in support groups reported feeling less psychological distress and physical pain than those who did not. Other studies have shown similar results among people with other diseases. And in my clinical work I have seen patient after patient feel better and experience less severe symptoms after participating in group-based programs.

Why do groups help? They offer support, compassion, friendship, and understanding. But for infertile women, they give so much more: verification that your feelings are natural (or, as one patient puts it, "It justified that my feelings were normal and I wasn't psychotic after all!"), advice for coping with difficult situations, tips on enduring infertility treatment, and firsthand descriptions of what it's like, say, to give yourself a shot or to ask a manager for a medical leave of absence or to confront a doctor about a disagreement over your treatment plan. Group support is particularly helpful when you're about to make a difficult decision, when you're exploring new parenting options, and when you're coping with a loss.

I highly recommend that you find an infertility support group to join. The first place to look is Resolve, Inc., a nonprofit infertility organization. Resolve has local chapters in every state that hold seminars and lectures and that sponsor professionally led support groups (see the Resources appendix for contact info). If there are no Resolve groups near you, hit the phones and try to find another organization that sponsors them. Ask your doctor or check with your local hospital, infertility clinic, or women's health center to see if the staff knows of any organization that runs infertility support groups.

If all else fails, start your own informal group. Although it's great when a support group has a trained leader to facilitate discussion, an informal, leaderless group is better than no group at all. To start your own group, ask your doctor to post a notice in his or her office saying that you're looking for other women who might like to get together once a week or so for decaf. I bet you'll get a lot of calls.

CHAPTER THREE

Coping When Everyone But You Has a Baby

My patient Nina and her husband, Mike, tried at first to conceive naturally. Later, as Nina underwent medical treatments, she began to feel overwhelming rage whenever she saw a pregnant woman.

> *I'm not ordinarily an angry person. I felt that I was being excluded from the club, and I was missing out on what life was about. I was so, so angry. I hated everyone who was pregnant. If I saw a pregnant woman walking down the street, I would curse her in my head.*

Eventually Nina became pregnant through IVF, and she is now the mother of a healthy one-year-old. But she still remembers those bitter feelings, and she hasn't completely let go of them, even though she has a child.

> *Whenever I hear of someone getting pregnant easily, I get upset. I want to tell them, "You have no idea how lucky you are that you didn't have to go through the hell that we've been through."*

June, who endured multiple miscarriages, felt overwhelming sadness.

> *After my second miscarriage, I started to think that I was never going to have a baby. That's when I started to see only pregnant women on the street. Every one of our friends was having a baby, talking about babies, having baby showers, and it was just a nightmare. There are people who are not our friends anymore because they didn't understand what we were going through, and it was too hard for me to see all those babies. Seeing people in a state that you want so desperately to be in is so hard. You feel sorry for yourself, you feel jealous, and you feel hurt.*

Laura felt that no one understood her situation.

> *We had pretty much excluded everybody from our lives who was pregnant or who had a baby because I couldn't stand it. That's all I heard, all I saw anywhere. People constantly asked, "When are you going to have a baby? Why aren't you pregnant? Why don't you get with the program?" I spent so much time crying. I told them various things, from "We're working on it" to "It's not that easy for us," or else I'd burst into tears and not say anything. Close friends knew what was going on, but if they were having kids themselves, they were just not that sympathetic. I couldn't believe some of the things people would say. One person I worked with knew that we were going through all of this. After his wife had a baby, he called me up one day and said, "Why don't you come over to dinner so you can see what it's like to be a parent?" I have never spoken to him since.*

Ann felt exceedingly jealous of pregnant women while she was going through infertility treatments.

> *My sister-in-law announced her pregnancy early last summer. I managed to avoid her and my brother for some time, but then I found out that they were going to visit my parents at their beach house the same week my husband and I had planned to go. Usually they go later in the season, but they decided to go early because of the pregnancy. This infu-*

riated me. I had an IVF scheduled a few weeks before the trip, and if it didn't work, I had no intention of spending a week with the expectant couple. So I made reservations for my husband and me to go to Florida instead. I decided that if the IVF worked and I got pregnant, we would go to Maine, and if it failed, we'd go to Florida.

The feelings experienced by Bella, June, Laura, and Ann may sound extreme, but they're not unusual. In fact, they are quite typical of many of my infertility patients, who say they feel more jealous, angry, resentful, spiteful, and hostile than they ever have before. When you think about it, it's not too surprising that infertile women are angry. Research shows that nearly half of infertile women consider infertility to be the most upsetting experience of their lives. It makes sense, then, that such an unsettling experience would trigger some of the most negative emotions of their lives. What is difficult for infertile women to accept, however, is that many of these troubling emotions are aimed squarely at people they love—their fertile sisters, neighbors, friends, and co-workers.

One thing you need to remember when you feel hurt/disappointed/angry with your friends is that people who haven't gone through infertility may not know what to do or say. What may sound horribly insensitive to you might be their clumsy attempt to try to help you feel better. A lot of people are ignorant of what infertility entails, and if you look at how the media portray infertility, with miracle twins born to fifty-something actresses, it's easy to understand how most people think that treatments are always successful. Also, others may not even know that infertility is an issue in your life. I'm sure that most of us have inadvertently hurt someone else through ignorance—I know I've done it. A while back, friends of mine hadn't gotten pregnant after five years of marriage. I wasn't in the field yet (while my friends dated and married, I was toiling in graduate school), and I remember repeatedly asking them when they were going to make me an honorary aunt. I'm now horrified at my insensitivity, but I didn't know any better then.

Also, when you're going through infertility, it's hard to imagine that friends don't get it. How can you be in such a crisis and have the people who love you fail so dismally to do and say the right thing? Think about it. Do you always do the right thing for everyone in your life? Are you always sensitive to crises in others' lives?

It can be difficult to remember that your fertile friends have real crises, too. Maureen, a patient of mine, was bitterly angry at one of her friends. A close friend who had two children seemed insensitive when Maureen was describing how devastating her unsuccessful IVF cycle had been. Maureen happened to be extremely wealthy, lived a fantasy lifestyle, and used to fascinate me with her stories of weekends in Paris, shopping trips at the most exclusive stores, and dinners at the finest restaurants. When I asked Maureen about her friend's financial status, she told me that in fact the friend had serious financial problems and was very worried about money. I asked Maureen if it was possible that hearing about Maureen's lifestyle was as painful for her friend as it was for Maureen to hear about the friend's children. She was stunned; that thought had never occurred to her. After thinking about it, Maureen realized that indeed the friend frequently changed the subject when Maureen's vacation plans came up.

Another patient of mine, Liv, came to her group session and said that she hated all fertile people. She told the group that on Sunday afternoon, on their way to the movies, she and her husband had dropped in on her best friend, who was home alone with her two small children. Apparently the friend had been rather rude. I asked Liv if perhaps the friend was jealous of their ability to saunter off to a movie on a Sunday afternoon, just as Liv was jealous of her friend's fertility. Indeed, when Liv called her friend and asked about it, the friend did admit that she had been frantic after being alone with two kids on a rainy weekend and was annoyed at Liv's ability to have so much free time.

Despite the realization that your friends and family have problems of their own, you may have friends or family members who need to be out of your life right now. For whatever reason, they are not able to be supportive of you during this crisis. You need to think of yourself and *your* needs. If someone is constantly making unsupportive or even cruel comments, you need to protect yourself. It may be only temporary; you may have good friends who are just hopeless whenever there is anything remotely medical in anyone's life, or perhaps they're going through a crisis of their own at this point and feel that infertility is nothing compared to their problem. Or maybe your single friends feel that when you find a man, you forfeit your right to complain about anything ever again, that wanting a baby as well feels selfish. These good friends will probably be back in your life after infertility, if you can forgive them and move on.

That's what happened with Melissa, a patient who experienced a late miscarriage after years of trying. She told a single friend how depressed and hopeless she felt, and the friend basically told her to stop being so weepy and to get over it. Melissa stopped communicating with her friend. Two years later she received a letter from the friend, who had gotten married and subsequently suffered two miscarriages. The friend wrote a very apologetic and remorseful letter to Melissa, saying that she finally understood what Melissa went through when she miscarried, and that she felt awful about her comments. The two of them were able to become friends once again.

Sometimes, however, infertility may help you figure out who your true friends really are. I've had dozens of patients whose infertility allowed them to see for the first time that some of the people in their lives were a negative influence. Let's face it, someone who lived next door to you while growing up, married a sarcastic jerk, had three kids in four years, and is constantly telling you how lucky you are *not* to have kids may not be someone you want in your life forever anyway. You don't need to make decisions now about whom you will forgive after infertility. Just think about who can support you during this crisis and seek out the company of such people.

If you have been infertile for some time, you may find yourself feeling some pretty intense emotions. See if these scenarios sound familiar:

You're horribly jealous. You may have been fairly content with your lot in life. Sure, there are other women who are prettier or thinner or who make more money or have better jobs or handsomer husbands. You've never begrudged them their good luck, and you've always been able to feel happy for them. Now, however, your equanimity has disappeared. You feel terribly jealous when other women get pregnant, you covet their babies, and you feel ferociously angry that so many other women have something you want so much but can't have.

Other pregnancies hurt. Hearing about another woman's pregnancy can set you off on a crying jag that lasts for days. All you can think about is the fact that she's pregnant and you're not. You may become so upset and depressed that you have to miss work or avoid contact with other people. Then, even after you regain your composure somewhat, your jealousy remains, twenty-four hours a day, seven days a week. You begin to forget what it feels like *not* to be jealous, and you start to wonder if you need professional help to overcome your feelings.

This is one of the toughest issues that my patients face. You love your friends, and you want all the best for them. When you find yourself getting angry at them for having a baby, you feel like such a bitch! Then you feel guilty and ashamed. It's just awful.

"Undeserved" pregnancies are unbearable. You get especially crazy when you hear that someone has gotten pregnant easily or accidentally. You feel that she doesn't deserve a baby, because she hasn't had to suffer. You're even more upset when pregnant women complain about their aches and pains. How dare they gripe about the minor discomforts of pregnancy when you can't even *get* pregnant! You'd give anything to have their indigestion or their varicose veins or their weight gain.

You avoid friends who are "trying." If you have a friend that you know is trying to conceive, you avoid her, fearful that any phone call or visit may bring news of an impending pregnancy. If she calls and leaves a message, you don't call her back, and when other friends mention her in conversation, you change the subject, because you don't want to hear that she's expecting. You even avoid friends with only one small child because of the possibility that they may be trying again.

You steer clear of social events. At first you pass up baby showers, but over time you start skipping picnics, cocktail parties, dinner parties, and family reunions, too. You just can't bear the thought that someone might announce a pregnancy. And you know that if she does, you'll fall apart and perhaps even make a scene. You'd rather be home alone.

Not knowing hurts almost as much as knowing. You hear from a friend that another friend is pregnant. This infuriates you. You're upset not only because your friend is pregnant but also because she didn't tell you the news herself. Of course, you haven't been picking up the phone . . . but still, you're doubly angry, even if it doesn't make sense.

Your agony is straining your marriage. Your jealous feelings and avoidance of friends and social events is enraging your husband, which in turn is enraging you. He says you're being unreasonable; you think he's being cold, and you don't understand why he doesn't feel as intensely as you do—after all, he wants a baby, too. Why doesn't *he* get so upset when other women announce their pregnancies? Most men just don't seem to report the intense feelings of jealousy that women feel in this situation. If your husband doesn't understand your jealousy, that can lead to conflict and arguments, as it did for my patient Maura and her husband Pete.

> *Pete just doesn't get it. He doesn't get why it upsets me so much, which causes enormous friction between us. I do not want to be around our pregnant friends, and I become hysterical when someone announces a pregnancy. And Pete doesn't understand it. To try to explain it to him, I used this analogy: We are saving for a house, but we can't afford it. I tell him, "Picture it like this: Even after all this time, we still can't afford the house we want. How would you feel if, while we're scrimping and saving, all of a sudden every one of our friends was handed a house for free? Absolutely free. Wouldn't that feel unfair?" When I explain it that way, I think he gets it, but then a few days later I realize he doesn't. But that's how it feels, and I don't understand why he can't understand it. You want something more than everyone you know, you're doing everything within your power to get it, you're sacrificing everything you have—time, money, energy, your body, your career—and you still aren't getting what everyone else is handed. It feels really unfair.*

What bothers women going through infertility more than just about anything else is that they feel as though they have no control over their situation. No matter how hard they work to attain their goal of getting pregnant, they fail. For many infertile women, that has never happened before. I hear this from patients over and over. All their lives they've found that if they work hard enough for something, they'll get it. They studied diligently in high school and were accepted to the college of their choice. They pushed hard in their careers and have been rewarded with promotions and bonuses and job success. They saved and sacrificed to buy the houses they wanted. Many have started successful businesses or run large corporations. But infertility marks the first time in their lives that working as intensely as they possibly can doesn't give them what they want. "All my life I've felt that if I put enough effort into something, it will happen," says my patient Stephanie. "In my job, in my relationships, in my family, in my education—I've been successful doing whatever I wanted to do if I put enough effort into it. This you have absolutely no control over. That drives you absolutely crazy."

Most infertile women do everything they're supposed to do. They follow their doctors' orders to the letter. They stop any behaviors, such as smoking, that might be hampering their fertility, and they do anything they can think of to be healthier: exercise, take vitamins, eat right, sometimes even quit stressful jobs to focus completely on getting pregnant. But they don't get pregnant. It doesn't make sense—a concept that psychologists refer to as cognitive dissonance—and it's terribly difficult to accept. What makes it even harder for infertile women is that as they struggle, the people around them—sisters, friends, co-workers—get pregnant effortlessly, sometimes even accidentally, like the goof-off kid in school who skips class, doesn't bother to study, and then aces an exam while the studious kid gets a C-minus.

Feeling angry and jealous is normal. But there are ways to stop negative feelings from taking over your life. And there are ways to manage your life so that you can avoid the pain of being around pregnant women—without cutting yourself off from the crucial social support that can help you through your infertility. If you're choosing to tell your friends what you're experiencing, here are some ways to help them help you.

 Create a How-to-Tell-Me Plan

Hearing that friends are pregnant can be terribly upsetting to most infertile women. Not only must you cope with the fact that someone else is expecting and you're not—again!—but also you often have to deal with this information without warning and in public, since the happy couples have a habit of springing their news at a party or holiday gathering. And you have to pretend to be elated about it! This is so hard.

You can't stop your friends from getting pregnant, but as difficult as it may seem, you can ask your close friends to be sensitive to your feelings when they make their announcement. Start by figuring out how you would like to be told. For example, it's agonizing for most infertile women to hear pregnancy news from someone on the phone or in person, because they immediately have to gush excitement and pretend to be joyfully delighted even when they're dying inside. That is an extremely painful process. So some of my patients ask their good friends, "When you find out you're pregnant, would you mind writing me a letter or sending me an e-mail?" Or "Please tell my husband, and he'll break the news to me." Or "Call me at home when I'm not there and leave a message, so I can process it privately, and I'll call you when I feel ready, after I've gotten over it. I need to hear the news in private, cry in private, and have some time to cope with my own sadness. Then I can be happy for you." You can also ask them, if they plan on making a surprise announcement at a social gathering, to tell you ahead of time so you can either steel yourself for it or skip the event.

You have to be careful how you discuss this with your friends, because you don't want to hurt them. If talking about this in person feels awkward, write a letter. Say something like "It's not that I don't want you to be pregnant. It's just that pregnancy announcements remind me of the agonizing experience I'm going through right now. You need to give me some time. Once I pull myself together, I can be happy for you."

Here's how someone in Lori's life handled it:

My best friend recently got pregnant with her second child. She knows about my infertility, and she didn't want to call and tell me she was pregnant. She realized how hard it would be for me. And it was, when she finally did tell me. But just the fact that it was hard for her to tell me made it better for me. It meant so much to me that she was so concerned.

Unfortunately, no matter how clear you are with your friends about what you need, some won't respect your wishes. They're not trying to hurt you. They're just so caught up in their own joy about being pregnant that they can't fathom why everyone else wouldn't be joyful with them. When that happens, try to forgive them and if necessary, avoid them for a while. You can control your own feelings and behavior, but not someone else's.

Plan Mental Escapes

Sometimes it helps to take your mind away from the hurt you feel after a pregnancy announcement, even for just a short time. Interrupting your painful emotions and negative feelings and focusing on something that is less emotionally charged can soften your body's physical reaction to stress. Mental escapes such as mini-relaxations, longer relaxation exercises, and guided imagery can slow your racing mind and give your body a break from physiological responses such as increased heart rate, blood pressure, muscle tension, and levels of stress hormones. What's more, many women find they can return from a temporary mental escape with improved perspective and a more detached attitude.

Mini-relaxations work well when your friend says, "Guess what? I'm pregnant!" Immediately after she makes her announcement, plaster a big smile on your face, ask her to tell you the details, and then, as she prattles on and on, do a mini. Slowly take a deep breath. Count one, two, three, four. Pause, then exhale. Four, three, two one. Repeat this mini-relaxation a few times—if you need to distract your friend from what you're doing, ask her what baby names she has in mind, and as she goes on about how she loves the name Gideon but her husband despises it, do a few more minis. Remember, the goal of a mini is to stabilize yourself for the short run, just as a doctor applies pressure to a wound that's gushing blood to keep the patient from bleeding to death. Only after the initial emergency is over does the physician consider such reparative measures as stitching the wound closed. Doing a mini will help you not to fall apart in front of your friend and anyone else who's present for the announcement, and it will aid you in maintaining your composure long enough to congratulate your friend on her good fortune ("Oh, you got pregnant on the first try? How nice for you!"), say goodbye, get in your car, drive calmly away, and then park on the next street and fall apart.

Later on, a longer relaxation session will allow you to elicit the relaxation response. Doing this will calm you and help you find your inner strength. It will also make you better able to perform some of the other coping skills, such as journal writing, that have previously helped you survive situations like this. Any kind of relaxation method will help—listening to a relaxation tape, breath focus, body scan, progressive muscle relaxation, meditation. After twenty to thirty minutes of relaxation, you're likely to find yourself refreshed and better able to cope with your friend's news.

Many of my patients find guided imagery to be a particularly effective mental escape after a friend announces her pregnancy. Not only does guided imagery elicit the relaxation response and help you flee a situation that is causing you mental anguish, it also permits you to become totally absorbed in the sensual aspects of a beautiful, safe haven where you feel comfortable, at home, and at peace (and where there are no pregnant women). You can use guided imagery to take yourself to a sandy beach, a mountain waterfall, a field of flowers, or even the house where you visited your grandmother as a child—whatever offers a peaceful, safe emotional respite for you. You can direct yourself to your safe haven or ask your husband or a close friend to verbally steer you to your destination. Do remember, though, that guided imagery works best when it is practiced first in peaceful situations, so that when you use it in a stressful time, you're returning to a sensory scene that you have already carefully created.

 Draw Up a Coping List

After a while you'll start to find that certain activities help you cope when a friend announces her pregnancy. These coping strategies vary from woman to woman, and what works for one may be worthless to another. As you experiment with different strategies, make a list of what works. Have an open mind, and try lots of different approaches. Think about what has helped you protect yourself during other life crises—maybe those tactics will help in this situation. Ask yourself this: What helps me more, to distract myself or to talk about it? To dwell on it or to push it out of my mind? Knowing the answers is important, because if distraction is what you need, then spending an hour on the phone complaining to your mother will only make you feel worse.

Make a list of ways to help yourself cope. Call it your "How to Cope When a Friend Gets Pregnant" list. Keep a copy of it on the refrigerator, next to the phone, or in your purse, so you can quickly see it when you need it. Then, when you're confronted with an announcement, grab your list and do something that will make you feel better or distract you from your thoughts. Here are some things that my patients keep on their coping lists:

- **Go for a mindful walk.** Focusing on what you see, hear, smell, and touch can take your mind off your friend's pregnancy.
- **Call a buddy in your infertility support group.** She'll understand how you feel better than just about anyone else will. Or call a friend or family member who has supported you through other difficult infertility moments.
- **Go to a movie.** Few things are as distracting as a great flick.
- **Get away for a few days.** If you can manage it, removing yourself from your everyday surroundings may help.
- **Write in your journal.** Writing has been shown to reduce stress and anxiety, and it helps you vent your angry feelings without hurting others.

- **Write a letter.** If you feel as though your friend handled her announcement badly, and you feel angry with her, write her a letter. Feel free to pour out your feelings, explaining why you're hurting and how you wish she had told you. But don't mail the letter—not right away, at least. After you finish writing, set the letter aside, read it again in a few days, and then decide whether you want to mail it as is, rewrite it, or just throw it away.
- **Talk to your partner.** If he is compassionate, he can be a great help. He can be even more effective if you talk beforehand to help him understand what he can say to help you. If your husband is terrible at comforting you in situations like this, however, opt instead to take yourself out for dinner and a movie with girlfriends.
- **Take it out on a pillow.** If releasing your anger in a physical way helps, punch a pillow or shut the windows and scream. You'll let out lots of angry energy, and nobody else will know.
- **Work it out.** Some women find that immersing themselves in a project helps, because they distract themselves and accomplish a task at the same time. If you're this type, scrub a floor, mow the lawn, or do some other satisfying chore.
- **Be artsy.** Creating something beautiful—painting a picture, writing poetry, composing a song, sewing a quilt, or making a collage representing what they would like in life besides a baby—can help some women replace feelings of sorrow with joy. Other women find solace in creating angry art, such as a dissonant musical selection or a disturbing painting. Remember, you're doing this for yourself as a way to ease your pain, and nobody else has to see or hear your work.
- **Pray.** Attending a service, saying prayers, focusing on a religious word or phrase that brings you peace (such as "Lord have mercy" or "shalom"), or reading the Bible may comfort you. Keep in mind, though, that if you are angry with God and feel that your infertility is a punishment, praying might only make matters worse.

- **Strike a pose.** Yoga can help remove your attention from your racing mind by focusing instead on doing something wonderful for your body.

Think About Your Thoughts

Cognitive restructuring is the process of examining your thoughts, inspecting them for truthfulness, and realigning them to be more accurate and less destructive. As we discussed in Chapter 2, all of us have certain recurring thoughts or phrases that we say to ourselves repeatedly and unconsciously, almost like tape loops that play over and over in our heads. For women struggling with infertility, the most common recurring thought is "I'm never going to have a baby!" This is what most infertile women think automatically when they receive bad test results, when they get their period, or when treatments fail. And it is, in my experience, the first thought that most women have when they hear the words "Guess what? I'm pregnant!" Before the jealousy, before the anger, before the sadness, comes the thought that no matter what you do, you'll never have a baby.

With cognitive restructuring you can put the "I'll never have a baby" tape loop to the test. Using the four basic questions that are the core of cognitive restructuring, you can free yourself from the pain that this negative tape loop brings you. Let's ask our four cognitive restructuring questions about the statement "I'll never have a baby."

Question 1: Does this thought contribute to my stress?
Answer: You bet it does. It's an all-or-nothing thought that robs you of hope and leaves you feeling that no matter how hard you try, you'll never achieve your goal.

Question 2: Where did I learn this thought?
Answer: Has someone been telling you this—your mother-in-law, perhaps? Your husband? Or have you just created it for yourself? Do you lack confidence in your doctors and feel subconsciously that they aren't

doing enough for you? Search deeply for the source of this thought so that you can face it and defuse its power.

Question 3: Is this thought logical?
Answer: No, it's not. You're doing everything in your power to have a baby. You're following your doctor's orders fully. You've got a procedure scheduled for next month, and you wouldn't be undergoing infertility treatment if your doctor thought there was no chance that you'd conceive. If you're feeling this way because you've had a string of failed procedures, resolve to talk with your doctor about moving on to the next level of treatment. You may be tempted to think, "Yes, it is true that I'll never have a baby because I've been trying for five years and I'm still not pregnant!" In that case, remind yourself that you have a Plan B (you should always have a Plan B, whether it be the next level of treatment, gamete donation, or adoption). Maybe the fact that you're feeling this way means it's time to move on to Plan B.

Question 4: Is this thought true?
Answer: No. Even in the most extreme case—you've had a hysterectomy and your husband has a sperm count of zero—you can still have a baby. Perhaps you can't conceive a biological child, but you can adopt, or have a surrogate carry a child conceived with donated egg and sperm. Even couples with no chance whatsoever of conceiving can still become parents.

By identifying and restructuring your negative thought, you can deflate its ability to hurt you. Over time you can teach yourself to replace the thought "I'm never going to have a baby" with "Somehow, one way or another, I'm going to be a mother." And eventually when a friend says, "Guess what? I'm pregnant!" you'll be able to tell yourself, "I'm sad that she is pregnant and I'm not. But I am doing everything I can to become a mother, and eventually I will succeed. It may not be as easy for me as it was for my friend, but I will succeed."

You can do a mini version of cognitive restructuring also. This is handy when you're in the middle of something else and need to quickly stop a negative thought from derailing you, but you don't have the mental focus to subject your thought to the four questions. It's especially helpful when you see pregnant women or babies while you're out and about—say, at the grocery store or the mall. Just ask yourself Question 3: "Is this thought logical?"

Imagine you're standing in line at the store and a pregnant couple gets in line behind you. Your first reaction is "I wish that were me." Try immediately to identify your thought and subject it to the test of logic—and try to have some fun with it, too, since humor is an excellent way to pull yourself out of a stressful situation.

Question: Is this thought logical?
Answer: Of course not. I don't want that baby. I want my own baby. If I wanted someone else's baby, I'd adopt. And if I adopted that couple's baby, it would have the father's big nose and the mother's weird hair. Yes, I want to be a mother, but, no, I don't want their baby. I'm doing everything I can to have my own child, and when I do, it will have a beautiful nose and just the right hair.

This doesn't work for everyone, but as I always tell my patients, it's worth a try. The most important point to take out of this is that when another woman announces a pregnancy, it is not the pregnancy in itself that is hurting you but the thoughts in your own mind that are causing anger, jealousy, and pain. If you can find ways to restructure your thoughts and reactions, if you can gain control of the negative tape loops in your brain, you can take some of the sting out of other women's fertility.

 Work to Get Over It

It's possible that nothing but a few days of tears will get you through a pregnancy announcement. And that's okay—you have to give yourself permission to feel sad. If you find yourself feeling lousy much longer than three or four days, however, or if your sadness interferes with your everyday life, consider talking with a therapist about the possibility that you're depressed.

You may just have to weather the storm. But as you do so, keep this in mind: Infertility is a temporary crisis, and it is not going to have long-term repercussions on your emotional state. No matter how crappy you feel today, you have to remember that you will survive this, no matter how your infertility turns out. Right now you may feel the worst you've ever felt in your entire life. And when your friends get pregnant, you may feel even worse. But a few years from now you'll be okay.

How do I know? First of all, I've seen how my patients rebound. When I first meet them, they're at the end of their rope. But as they go through my program, they improve. And when I talk to them months or years later, as I often do, they're fine. In fact, many of them tell me they are stronger, their marriages are more robust, and their lives are more satisfying than they were before infertility—*whether or not they've had a biological child.* That doesn't mean they're glad they were infertile. But they survive it, and most of the time they emerge stronger.

There's even a study that backs this up. A few years ago Canadian researchers looked at women in their sixties and seventies who had struggled with infertility when they were younger. A third of the women had had biological children, a third had adopted, and a third had remained child free. The researchers examined the women and found that their psychological profiles were identical. They were all equally happy. And they had survived their infertility just fine.

If you're angry and jealous and resentful of other women, don't be ashamed. You were probably brought up to believe that jealousy is terrible, that coveting what someone else has is a sin. You may tell yourself, "How dare I feel so horrible for myself when friends are experiencing such happy news? What a bad person I am not to share in their joy!" My response to that is this: Of course you're not ecstatic when someone else so easily gets what you want so much. It's human nature to feel jealous and to get upset in the face of injustice. You're doing a perfectly normal thing—you're reacting negatively to an unfair situation. There's no doubt about it: Infertility is unfair. Infertility is unjust. You *should* be feeling angry! Every woman who is dealing with infertility feels angry.

I'm not worried about you because you feel this way—but I'd worry if you didn't. *That* wouldn't be normal.

Here's what I was told by my patient Erin, who's been unable to conceive for several years despite three IUIs and is about to undergo her first IVF:

> *I asked my therapist how to deal with other people getting pregnant. I asked him, "What am I supposed to do? I need a way to handle it." And he looked at me and said, "There is no good way to handle it. You're going to be upset. And that's okay. That's what's supposed to happen. You're not supposed to be OK." I was trying to figure out how to dodge it and not feel the feelings, but he said, "You have to go through it. It's part of the whole process. It doesn't make you a bad person if you're upset that your friend is pregnant and you're not. You just have to feel the feelings and deal with them. If you ignore them or try to act like they're not there, then you're not working through it; you're putting it off." So now if one of my girlfriends calls and says she's pregnant, I get upset, I deal with it, and then eventually I can be happy for her.*

Deirdre, who's been infertile for three years, says this:

> *I work with a lot of young women, and every other week someone is announcing that she's pregnant. There have been several occasions when I was really upset, but lately it hasn't been too bad. When it happens, you say to yourself, "Why her? Why isn't it me?" But you can't go on like that forever. Now when it happens, it upsets me a little, but not like in the past. I've made a lot of progress with this. I tell myself that it's going to work out eventually. Something is going to work out somehow.*

CHAPTER FOUR

You and Your Husband

When Ashley and Ben began trying to conceive a child four years ago, they thought of it as a thrilling journey. "We were both really excited," says Ashley, who is now thirty-four. "It wasn't just sex at the time of ovulation, it was all the time—we were like honeymooners again." A year later Ashley wasn't pregnant yet, but she and Ben didn't feel too discouraged—after all, he traveled a lot for his job, and they figured they just needed to pay closer attention to the timing of her cycle. They still considered conception an adventure, and they felt very close, both emotionally and physically. "We consoled each other, our communication was good, we were sharing—we weren't experiencing any grief," Ashley recalls. "We were disappointed, but that was as far as it went."

Tests showed that although Ashley and Ben had two strikes against them—he had a low sperm-motility rate, and her progesterone levels were low during one phase of her menstrual cycle—they probably could conceive naturally, particularly if Ashley took medication to boost her progesterone production. They were somewhat disheartened when they discovered these biological disadvantages, but their relationship remained strong. "We were in it together," Ashley says. "We were there for one another, and we were able to talk about it easily. We were still optimistic and hopeful."

During the second year of trying to get pregnant, however, things began to unravel. "It became sex on a calendar," Ashley says. "It got to the point where that was the only time we did it." They saw an infertility specialist and tried several IUIs but, still, no baby. "The load to bear got heavier and heavier," Ashley says. "The whole 'we're on an adventure' thing started turning sour." Ashley felt depressed and became upset easily. Ben held in his grief, afraid to share it with his wife. "He wanted to grieve for himself and express his sadness and disappointment, but at the same time he didn't feel he could do that in front of me, because I was just as devastated, if not more so. He kind of retreated by pulling into himself and not really talking about it. And all I wanted to do was grieve constantly—I had no motivation to do anything else. I was impossible to deal with. And we existed that way for a good two years."

Most couples are completely unprepared for the strain that infertility places on their relationships. Who could possibly be equipped to deal with the feelings of disappointment, the interminable medical treatments, and the dashing of hopes and dreams that come with infertility? The biggest problems many couples faced before infertility were issues such as where to seat Aunt Ethel at the wedding reception and whether to go with wall-to-wall or hardwood in the living room. But the difficulties encountered while dating, planning a wedding, and setting up a home seem insignificant compared to the emotional roller coaster of trying—and failing—to conceive. Infertility is the first crisis that many husbands and wives face together, and unless you learn some new coping mechanisms, it can crack your marriage wide open.

Marriage therapists say that sex and money are the two top causes of marital strife, and with infertility you're likely to struggle with both. Problems often start in the bedroom. The carefree, playful sex that you had when you were first married can turn into a regimented chore when you're trying to conceive. At first, sex without birth control is a blast, because you combine the excitement of unprotected sex with the joy of thinking about starting a family. But after a while the excitement fades. Month after month after month the act fails to produce what it's supposed to. Lovemaking gets cluttered up with ovulation-predictor kits, doctors, schedules. Often women start to want to have sex only in midcycle, because that gives them the greatest chance of having a baby. And men start to feel that the only reason their wives want to have sex with them is to make a baby, that they're not interested the rest of the month. Both start to associate sex with failure. Men feel that their wives view them as just sperm-producing machines. Sometimes anxiety causes men to become impotent, particularly during the middle of the cycle, when their performance is most necessary. All the intimacy, emotional bonding, sensuality, and spontaneity go right out the window—sex must be performed not when you feel amorous but on demand, when the calendar or the doctor says it's time. A nonstop barrage of shots, blood tests, and surgical procedures can leave you feeling so crummy that you don't even want to hold hands, let alone make love. And let's face it—if the infertility is your body's fault, you may feel as though you're letting your spouse down because you can't make a baby. Those feelings of inadequacy can interfere with desire and performance in the bedroom. Making love is one of the best ways for many couples to connect emotionally, and when their sex life is associated with failure, frustration, anger, and resentment, a crucial avenue of emotional expression disappears.

Then there's money. Procedures such as IVF can cost as much as a small car, and few people have health insurance that adequately covers the procedure time after time after time. Indeed, only a few states require insurance companies to pay for infertility treatments. If you're not covered, you may be writing some pretty big checks and having some pretty big arguments with your spouse over how much to spend—and when to stop spending. Do you borrow from family members to try IVF one more time? Do you apply for a second mortgage on the house to pay for a treatment that has only a small chance of success? What if one of you says yes and the other says no—what is the compromise?

Beyond sex and money, there are many other factors that may chip away at your relationship. As we discussed earlier, depression is a huge problem for women who can't conceive or carry a pregnancy to term. Fertility drugs can push the most even-tempered woman into roller-coaster-like mood swings. That was Ashley's experience.

> *I was so moody and depressed. Often I'd find myself getting really short tempered. If my husband asked me for something when I was in a bad mood—it didn't matter what it was—I was ready to go for the jugular. He would ask me to do a favor, and I would think, "Oh, my God, what does he want now? I'm always doing things for him." I was so nasty and gross to him—and this is the man I love!*

The typical symptoms of depression—moodiness, loss of interest in favorite activities, changes in appetite or weight, sleep problems, energy loss, feelings of worthlessness or inappropriate guilt, difficulty thinking or concentrating, and even recurrent thoughts of death or suicide—can really put pressure on a marriage. No matter how loving you may be, it's tough to connect with a spouse who is frequently withdrawn, sad, lethargic, and seems uninterested in you and in the things you both used to enjoy doing together.

Unfortunately, infertility demands so much from couples—learning how to give injections, figuring out ways to avoid baby showers, mastering as much health information as a first-year medical student—that it can be hard for you to think of anything else. So many conversations, decisions, and outings revolve around infertility that a couple may forget, or not have time, to focus on anything else but getting pregnant. Before you know it, all the fun drains from your marriage, and there's nothing left but the terrible stresses and demands of infertility.

And what makes it all worse is that women and men handle infertility differently.

There are several reasons for this. First, women tend to crave babies in a way that men don't. Blame it on evolution if you like, but the truth is, our ancient ancestors survived because women bore, raised, and protected children while men went out to slay dinner. To this day, even though our sensibilities and roles are far different than those of our ancestors, women are still biologically and neurochemically hardwired to want babies in ways that men simply are not. Yes, you and your spouse both may yearn for a baby, but you yearn differently. You both may grieve when conception fails or miscarriage occurs, but you grieve differently. You both may be depressed by infertility, but you're depressed differently.

Second, because the woman's body is the intended baby factory, women bear the brunt of the physical strains of infertility treatments. Even when the infertility is due to a male factor, it is the woman who undergoes most of the medical interventions. The woman's body is the infertility laboratory, and even the most loving, supportive husband in the world can't truly understand how deeply overwhelming it can be to have months of shots and blood tests and ultrasounds, to see that monthly trickle of menstrual blood, to undergo a D&C, to burst into tears at the sound of the word "adoption." Likewise, women don't truly understand the enormous shame a man who equates potency with manliness may feel when he can't make his wife pregnant, or the anxiety and embarrassment of having to fill a lab bottle with sperm for an IUI or IVF.

Third, because of the divergences in their emotional coping styles, men and women tend to respond quite differently to the burden of infertility. Sadly, those at-odds responses can cause major rifts in a marriage at a time when spouses need each other's love and support more than ever.

Many women complain that their husbands don't appear upset about infertility. When pressed, the men will own up to being disappointed that their wives can't get pregnant, but they don't seem to take the whole thing as personally as their wives do. For a woman, getting pregnant often becomes something of an obsession, and a failure of any kind, from a period to bad test results, can throw her into days of depression. But so many men seem to be able to take conception setbacks in stride, to express disappointment but then move on. Often a man will react to a negative pregnancy test no more emotionally than he does when he reaches for his favorite cereal in the morning and discovers that the box is empty. This apparent lack of emotional response can infuriate women and cause enormous resentment, arguments, and lots of hurt feelings.

Well, let me tell you what I've learned from counseling hundreds of infertile couples: Men do feel the pain of infertility. They do care. They do respond. But most tend to behave so differently from the way their wives do that their wives don't even see their pain. Most of the time, that is—some men can't or won't be supportive during this time, and if that is your case, make sure to read the section at the end of this chapter called "On the Other Hand." But most of the time, husbands do try to be supportive. Let me illustrate this by telling you about Bonnie and Dan. They had been trying to conceive for four years. For the last year their marriage had become shaky. When I met with them, Bonnie was miserable about her infertility. She talked about it frequently and would often lapse into tears when she discussed her latest failed medical procedure. Dan, on the other hand, seemed almost unaffected by the situation. Test results that would send Bonnie off into a teary bout of depression didn't seem to bother Dan; sure, he said, he was disappointed, but what good did hysterics do? After several sessions with Dan, however, I came to understand that although he didn't display much emotion and sometimes acted almost cavalierly about infertility, he was in fact enormously affected by the problems the couple was facing.

> *I'm a very practical person. I don't see what good it does to get upset about things. It's really hard for me that Bonnie gets so emotional about not being able to get pregnant. The worst part is that nothing I do to try to help her makes a difference. I just can't seem to make her feel better, and that's a hard thing for a husband, when you can't help your wife.*

Dan felt he was a failure as a husband because he couldn't lift his wife's spirits, and he dealt with that feeling of failure by shutting down and isolating himself from her. But what Dan needed to understand was that he wasn't responsible for his wife's sadness, and that he shouldn't blame himself for not being able to lift her out of her depression. At the same time, both Dan and Bonnie had to make peace with their partner's style of coping. Bonnie resented Dan because he could calmly talk about infertility one minute and happily watch a basketball game on TV the next. And Dan resented Bonnie because of the highly emotional way in which she responded to infertility. Once the couple acknowledged their very different coping styles simply by talking about the issue at our sessions—and gave each other the permission and space they needed to cope in their own ways—Bonnie and Dan began to repair the tears in their marriage.

Sometimes husbands shut down emotionally because they feel they have to be strong in order to support their wives. In these cases the husband feels that if he doesn't hunker down and become the emotional bedrock of the couple, chaos will result. That's what happened with Meg and Steve, in his words:

> *Honestly, after we went through a couple of unsuccessful IUIs, I felt like I was in an incredibly awkward position. I felt sad and disappointed, but at the same time I didn't feel I could grieve in front of Meg because I thought she'd lose it if I did. She was so upset all the time. I took on the role of tough guy, supporter. Meg called me "the rock." But I felt just as bad as she did. I just didn't show it. After a while Meg started accusing me of not caring, and that really hurt me, because I did care.*

Women often cope with the problems of infertility by withdrawing from their husbands or into the world of infertility, where there is always another book to read or vitamin to try or expert to consult. Men tend to cope by retreating from their wives, to the television or the office. Women may get so obsessed with infertility that they talk of nothing else. Men have less experience talking through their feelings, so they may say nothing at all. Women may become moody and sad. Men may try valiantly to coax their wives out of their depression, but nothing they say seems to help. Communication fails. Pain and resentment build. Spouses pull away from each other, not intentionally but because it's the only way they know to respond to each other's closed-off behavior and to survive the stress of infertility.

But there's hope. There are ways to solve the problems, to reconcile the differences, and to unblock the channels of communication. I've seen the damage that infertility can do to a relationship. But I've also seen couple after couple successfully use a variety of mind/body techniques to break down the barriers that separate them, to open their hearts to each other, and to rekindle the love, affection, and fun that infertility has stolen from them.

Staying in the Now

Perhaps you often find yourself thinking ahead to the future, dreaming of the day that you and your husband will caress your swollen belly, race to the hospital to deliver a baby, push a toddler in a swing—or even send a grown child off to college at your alma mater. Or maybe you're constantly obsessing about the past—why did you wait so long to try to conceive? Is it possible your promiscuous behavior before marriage contributed to your infertility? If only you had started trying when you were younger. . . . If your mind is stuck in the past, the future, or anywhere else but now, practicing mindfulness could go a long way toward helping to improve your relationship with your spouse.

Mindfulness is based on the principles of Tibetan Buddhism. It is a technique that centers your mind in the present, rather than letting it wander to the hopes of the future or the regrets of the past. Being mindful means appreciating what you have, rather than longing for what you don't have.

For infertile couples, mindfulness means moving your focus away from getting pregnant and concentrating instead on the other parts of your life that bring you joy—the wonderful way it feels to go for a walk hand in hand, the joy of working together to complete a satisfying home-improvement project, the fun of having an ice cream sundae together. It also means fully experiencing and appreciating even the most routine parts of your life together, from eating leftover pizza in front of the TV to kneeling beside each other at church.

Alana, thirty-four, has been trying to have a baby for three years. Following IUI treatment, she managed to conceive twice, but both pregnancies ended in miscarriage. After the first miscarriage Alana realized she was becoming unhinged. She felt depressed, she was snapping at her husband, she was losing interest in sex. "I realized I was just not coping well with this—I needed help. I knew there had to be some way for me to take more control over this process and to make myself feel better about it and be strong enough to move forward."

Alana enrolled in the mind/body program, and she used mindfulness as a way to stay aware of the fact that although she was terribly upset by her miscarriage, there was more to her life than her infertility. "Before I started the program, my attitude was really negative. Everything in my life seemed bad. But when I started to practice mindfulness, I realized that's not true at all. Just one aspect of my life—which, granted, is a big part of my life—is bad, but there are a lot of other wonderful things that I should be thankful for."

Alana realized that infertility had been making her miserable, which in turn had reduced the quality of her marriage. But mindfulness helped to change that. "I no longer feel like I am an unhappy person. I'm a happy person dealing with a huge challenge. Now I wake up in the morning and am mindful of the simple pleasures of life—enjoying a cup of coffee with the newspaper, going on a walk with a friend, reading a book. I tried to become amazingly aware of the things that people take for granted. And I slowly began to feel the burden of infertility not weighing so heavily on me and on the relationship with my husband." Alana also found herself better prepared to cope with her second miscarriage, which occurred after she finished the program. "I think I would be in a much worse place emotionally right now if I didn't have the class to fall back on. There's no doubt about that in my mind."

Mindfulness is being completely aware of every sensation—what you see, hear, smell, feel, and taste. Being mindful means living in the moment, thinking only of the way things feel right now, not about your worries of the past or your concerns for the future.

I advise infertile couples to begin each day mindfully. If your schedule allows it, try starting the day with a mindful walk, either alone or with your spouse. (If you can't manage this every day, try for weekends.) But this isn't a fitness walk. Instead of thinking about how much road you can cover and how fast you can get your heart to beat, walk mindfully. Enjoy how good it feels to be outside, how nice it feels for your body to be moving. If you're walking with your spouse, be aware of the rhythm of your walk, the sounds of your spouse's breath, the feel of your hands clasped together. There's no need to talk—just being mindful of the experience and focused on the moment is enough.

Be aware, though, that mindfulness can be a hard skill to learn. As you try to focus on the present, unwelcome thoughts are bound to intrude—for example, feelings of dread about your period being due next week—and interfere with a mindful walk. But when those thoughts enter your mind, acknowledge them and then let them go. Imagine putting them into a box that you'll open and inspect later, after your walk.

"I take mindful walks almost every day," Emma says. "I walk my dogs along abandoned railroad tracks near my house. The dogs run, and I walk slowly and mindfully. I find that now, even at work, I walk more slowly and mindfully. Before, I was always colliding with people when I'd turn corners at work. There's no need for that."

During the Sunday part of the mind/body program, when husbands are invited to participate, Emma and her husband, Vin, tried walking mindfully together for half an hour, and the experience was exhilarating, especially for Vin, because it gave them the time and the space to leave infertility behind and focus on being together in a quiet, loving way. "He was on a high for three or four days after that," Emma said. "He just loved it."

Starting your morning mindfully can be a wonderful way to center yourself for the entire day. But don't limit your mindfulness to morning. You can use mindfulness to increase your awareness and appreciation of your entire day, from the train commute in the morning to the time you spend chopping vegetables for dinner.

You can also enrich your relationship with your spouse by being mindful during all of the activities you do together—you can eat lunch mindfully, play golf together mindfully, even go to the movies mindfully. By focusing on what you're doing at the moment, you can help contain your infertility stress and prevent it from overtaking your relationship and your life.

I'll never forget a time a few years ago when I was teaching mindfulness to one of my mind/body groups. Their "homework" assignment was to do something mindful and report back on it to the rest of the class at the next session. When we met the next week, the women talked about what they had done mindfully—one had baked bread mindfully, another had taken a mindful shower, another had made a salad mindfully. When it was time for Sherry to talk about her mindful activity, she blushed deeply and told us all she'd had sex mindfully—and that she didn't realize she'd have to talk about it in class! We ribbed her mercilessly, asking for detail after detail. Eventually she overcame her embarrassment and told us that it was some of the best sex she'd ever had, and she intended to have mindful sex many times again in the future.

I still laugh when I think about it, but in fact Sherry was onto something. Mindfulness can be an incredibly useful way to bring the joy back to your sexual relationship, too. Sex becomes a task for many infertile couples, not a pleasure. The goal of infertile sex is often to get as much sperm as possible into the right place at the right time, which leaves little room for spontaneity and passion. There's not much you can do about that, but you can add mindfulness to your other romantic encounters. Remember what it was like having sex just for fun? Think back—fun sex is mindful sex, the kind you used to have when you first started making love with each other. To have sex mindfully, leave your thoughts of pregnancy and infertility outside the bedroom door and think instead about paying full attention to the sensual aspects of making love—the way your spouse's body feels, the way your body responds to your spouse's touch, the closeness you feel when you're all wrapped up together under a big, warm quilt on a cold winter morning. Be in the moment. Don't think about the pregnancy test next week or the procedure you had last week. Be in the here and now with your spouse, focusing deeply on what things feel like, what things smell like, what things taste like, what things sound like. Focus on your sensations and the intimacy that lovemaking can create.

If you're having trouble separating fun sex from procreating sex, and you have the extra space, you may want to experiment with using two different rooms—the guest room for trying to conceive, say, and your own bedroom for mindful sex. This setup allows you and your spouse to retreat to an infertility-free zone where pleasure and passion, not calendars and eggs and sperm, rule. Also, try taking breaks from the effort to get pregnant, as my patient Cara does.

"Infertility is very draining on your sex life," says Cara, who has been unable to conceive because of endometriosis. "There have been a lot of times within the past three years that I've said to my husband, 'Thanks,

but no thanks.' You're told when you can and when you can't. Your sex life becomes a part of your fertility process—it's no longer spontaneous."

To spark the passion in their sex life, Cara and her husband, Karl, decided to take occasional months off from trying to get pregnant. "We consciously agree not to talk about it. That makes you feel much freer." Cara and Karl have also held off on trying to get pregnant during vacations, which they consider to be infertility-free zones, with no drugs, no paying attention to cycles. "We are able to leave it all behind. We go hiking and biking and go to the beach and really just enjoy all those kinds of things."

Nonsexual Intimacy

Yoga can be incredibly refreshing for body and soul. In the mind/body infertility program, we teach couples yoga for a number of reasons. First, we advise our patients to stop exercising for three months to rule out the possibility that exercise is a factor in their infertility. (For a full discussion on this, see Appendix I.) A lot of the women are concerned about losing their body tone, so yoga is a safe way to keep them fit and toned. Second, yoga is an excellent form of relaxation. We teach a number of relaxation techniques, but for a lot of patients, yoga is the technique of choice to elicit the relaxation response.

Yoga is great for infertile individuals, but it can be equally helpful for couples. Couples yoga allows you to do something physical together that isn't sexual, so it doesn't bring up feelings of failure in terms of infertility—it allows you to comfort and nurture each other in a nonsexual way. A lot of couples touch each other only in a sexual context, and with infertility, often the only reason to have sex is to try to make a baby. Because of this, sex is frequently associated with failure—you're not even making love anymore, you're just trying desperately to get pregnant. For infertile couples, sex is anxiety provoking; yoga is anxiety reducing. It's a low-pressure way to do something sensual together. Couples yoga allows you to experience deep relaxation and physical intimacy. It's something you do together for your mutual benefit—and it feels good. While infertility can alienate spouses from each other, couples yoga can bring them back together. That's how it helped Marta and her husband.

I feel so much less sexual now than I did before infertility. But couples yoga helped my husband and me to get in touch again physically. It brings us physically close, which is great. Sometimes I have the natural instinct to push him away, when what I really need to do is allow myself to feel a lot more love, and to give love.

 "Finally, Someone Who Understands"

It may sound strange that having the support of other infertile women can strengthen a woman's relationship with her husband, but based on what I've seen within my program, group support can really help a marriage. Here's why: As we discussed before, try as they may, men just don't understand what it feels like to be a woman struggling with infertility. And try as they may, most men can't talk about infertility as openly and as constructively as women can. A woman who talks only to her husband and her fertile family members and friends has nobody who really, truly understands what she's going through. An infertility support group meets a woman's needs to communicate, to mourn, to share information, to bitch about medical treatments, even to laugh about the absurdity of it all. It makes her feel less isolated. And all that helps a marriage because if a woman's need to talk about infertility is met outside her marriage, she's less likely to rely so heavily on her husband. Getting group support is one of the best things you can do. Carolyn, whose infertility has caused something of a rift between her and her husband despite their efforts to stay close, knows firsthand.

My husband is an exceptionally sensitive, intuitive person, but he's not a woman. He doesn't quite understand what I'm going through. So his way of encouraging me or getting me to be strong doesn't always work. He just doesn't have the innate desire, the driving force to have children.

He doesn't know what it feels like to be denied that. On an intellectual level he does, but he can't feel it. It's the same with my girlfriends who've children and who have never had any trouble getting pregnant, and my sisters. They don't quite get it either. That's why these support groups are so helpful. It's a group of women who might not have become friends under ordinary circumstances but have become really close because they're the only other people who understand what you're going through. It creates an environment where you don't feel so alone.

Faith gave birth to a daughter two years ago after a successful IUI but has been unable to conceive again. She was a member of one of my mind/body groups, and the group continues to get together once a month.

With each failed cycle, my hope diminishes. There's part of me in this deep, deep pit—my husband and I had hoped to have six children—and no one can understand me but the women in this group. We know what we're all talking about. We have a bond, although it's a very painful bond.

A New Way of Thinking

Among the infertile woman's worst enemies are the automatic negative thoughts that pop into her mind, quite uninvited, and settle in to stay. These thoughts have little basis in fact; they twist other thoughts into an unrealistic mess; they plant anger, defensiveness, and fear in a woman's mind; and they can be a potential minefield in a marriage. Yet almost every woman struggling with infertility has them. *My husband is going to leave me and marry a fertile woman. My husband resents me because the infertility is my fault. Why do I have to go through all these treatments when the infertility is his fault?* These automatic thoughts cause nothing but trouble. With cognitive restructuring, you can identify your negative thought, closely examine it by determining its source and putting it to a test of logic, and then restructure it so it more closely represents the truth.

Let's look at one of the most common automatic thoughts that infertile women have—that their husbands are going to leave them for a fertile woman. Liz, a patient of mine, struggled with this thought. Here's how we restructured it:

> **Liz:** Frank comes from a big family, and he's desperate to have children. We can't conceive, and it's my fault. I'm afraid that Frank is going to leave me for a woman who can give him the large family he's always wanted.
>
> **A.D.:** So basically what you're saying is, if the infertility were Frank's fault, you'd leave him to find someone who's able to make sperm.
>
> **Liz:** Of course not! I didn't marry him for his sperm, I love him!
>
> **A.D.:** Then you're saying that your husband is a shallower person than you are?
>
> **Liz:** (defensively) He's not shallow. He's one of the most loving, committed, thoughtful people I've ever met.
>
> **A.D.:** Well, if he's so committed to you, why do you think he would leave you just because your reproductive system doesn't work? Do you really think he married you just for your eggs? Or because he loves you for who you are?
>
> **Liz:** You're right. He loves me for who I am. He may be disappointed that I can't conceive, but he's not the kind of person who would leave me for another woman because of it.

By looking critically at her thought and testing it for logic and truth, Liz saw that her fear was unfounded. Yes, her husband was disappointed that she couldn't conceive a baby, but by using cognitive restructuring and by working hard to communicate with Frank, Liz learned that Frank continued to love her deeply, despite *their* infertility.

Thinking about how your spouse would react if the situation were reversed can be a very helpful cognitive-restructuring tool. It's especially effective in the case of male-factor infertility. It's easy for a woman to feel angry when infertility is her husband's "fault" and yet she has to endure infertility treatments. It's easy to think automatically, "Why do I have to go through all these treatments when he's the one with the problem?" But by using cognitive restructuring and role reversal, a woman can realize that if her body were the cause of infertility and her husband were forced to undergo medical procedures in order to conceive, most men would agree to do so. By looking at it as a shared sacrifice that you'd both be willing to make, you can restructure negative thoughts and take away their power.

It's important to remember, however, that cognitive restructuring is not Pollyanna thinking. The point of cognitive restructuring is to dig out the truth, not to hide it behind sunny delusions. If your marriage is weak, cognitive restructuring, journaling, and other techniques may uncover those weaknesses. If you feel that as you search for ways to cope with infertility, you discover problems in your marriage, I urge you and your spouse to see a couple's counselor. Without realizing it, you may be trying to get pregnant to save your marriage, and if that's the case, your marriage may not withstand infertility (or a child, for that matter).

 The Relationship Quadrant

Knowing and avoiding their danger zones—the situations in which even the tiniest difficulty can escalate into a nuclear explosion—is incredibly important for an infertile couple trying to keep their marriage strong. But it can be hard for infertile couples to stake out those danger zones, because infertility puts them into entirely unfamiliar emotional territory. The you-left-the-top-off-the-toothpaste issues that used to cause problems pale in comparison to the concerns of infertility. How can you negotiate the newly formed hills and valleys in your relationship? I recommend a tool called the Relationship Quadrant.

Most of us have two general states of being—we're either OK or not OK. Depending on your personality, you may be impatient or relaxed, blue or happy, intolerant or tolerant. A woman may go back and forth between being OK with her infertility and not OK with her infertility. And your spouse has two states as well. Perhaps he's engaged or aloof.

Or maybe he's either calm or frazzled, strong or vulnerable. Whatever words you choose to use, the reality is that between the two of you, you have four potential interactions, or quadrants, that you can land in as a couple:

1. You're OK and he's OK, so everything's fine.
2. You're OK and he's not OK, which is all right, because you can help him.
3. You're not OK and he's OK, which is also all right, because he can help you.
4. You're not OK and he's not OK. This is the danger zone, a place you want to avoid.

To plot the relationship quadrant for your marriage, figure out your states—let's say you're either "calm" or "overwhelmed." Then determine your husband's states—let's call them "confident" or "insecure." Now think about what moves you from your OK quadrant to your not-OK quadrant, what moves you from being calm to being overwhelmed. This varies from person to person, but it could include having someone tell you she's pregnant, bad test results, delayed doctors appointments, getting your period. Maybe the things that move your husband into his not-OK quadrant have to do with his job—for a lot of men, their two states are influenced by how they're doing at work, so maybe you know that your husband's Monday-morning meeting with his boss or the end-of-the-month sales figures have the power to knock him into his not-OK quadrant. Once you have a list of things that push you both into your bad quadrants, you can start to develop prevention strategies that help you avoid that quadrant.

Prevention doesn't always work—there's no telling when a friend is going to announce a pregnancy, for instance—but it often does. Here's an example of where it works: Say you know that your period is due on Monday, the day he's usually in his insecure quadrant because of his weekly meeting with his boss. You know that when you get your period, you'll feel overwhelmed. Just being aware of this potential trip into the danger quadrant can help you avoid it. You'll understand why he's not fully available to comfort you when you get your period, and he'll understand why you may not be as able to help dispel his insecurities about his work. Knowing this can prevent the explosion that might occur when you look to him for sympathy and he doesn't provide it. You may even want to go one step further and shore up additional support for both of you—perhaps you can ask your mother or sister or an infertile friend to distract you with dinner and a movie, and he can go out for a beer and a game of darts with a friend who will be able to help boost his feelings about his career. Then, by Tuesday, you'll be feeling less overwhelmed, he'll be less insecure, and you will have avoided the arguments, the resentment, and the anger that can erupt in the danger zone.

He Listens, She Listens

You talk with your spouse, but is he really listening? Do you really listen when he talks to you? Pay attention to yourself the next time you're talking with someone. Usually when you have a conversation, you sort of listen, but you're also deciding whether you agree with what the other person is saying and thinking of a response. You're also, in the back of your mind, deciding what to make for dinner tonight, what you'll wear to work tomorrow, whether you'll be able to finish that proposal that's due Thursday—your mind goes in a million directions. In our multitasking society we very rarely just listen, and this natural tendency to half-listen can cause huge problems for an infertile couple.

I find that in almost every couple the woman wants to talk about infertility more than the man does. When I meet with couples, I feel like I'm in a shrink scene from *Annie Hall*—women say, "We never talk about infertility!" and men say, "All we ever talk about is infertility!" So what frequently happens is the woman talks and talks and talks and talks, and her husband doesn't listen. He shuts her off, and that leads to frustration on both sides. What women are looking for is to be heard.

This is where paired listening comes in. Paired listening gives each member of the couple time to talk without interruption and to listen completely. In our mind/body infertility program, we split up into groups of two, and then each person takes a turn talking for five minutes while the other one listens fully. The listener is not allowed to interrupt, argue, agree, or respond in any way—she just listens. At first it's hard for some women to just listen. They are tempted to interrupt. But once you get the hang of it, paired listening can be a fantastic communication tool.

Try this at home with your spouse. Talk about infertility via paired listening for ten minutes—five minutes a person—every evening. Often my patients find that those ten minutes is all they need. They can say what they need to say and then get on with the rest of their evening. Ten minutes of paired listening can be more effective than half an hour of fragmented conversation that is spent mostly on getting attention rather than communicating. Paired listening works because the wife is pleased to get her husband's undivided attention for a certain amount of time, and the husband is glad because he has to listen for only a limited amount of time. So many of my patients come back a week after learning paired listening and say it made them feel as though their husbands were really listening to them for the very first time.

It took paired listening for Tammy, who's been trying to get pregnant for four years, to realize how much grief her husband, Ed, felt about their infertility. Ed finally expressed some of his sadness, and Tammy helped him to cope with his grief.

> *He felt just as bad as I did, if not more so sometimes. But he didn't express it. He'd go into hiding—his coping was to disappear and play golf. And among friends and family a lot of the outpouring of support was directed straight at me and not at him. He kind of got pushed to the side. Now I'm much more careful to ask him how he's feeling and what he's thinking. I'm not just focused in on myself anymore. I'm looking at the relationship again, and we're a lot happier now than we were in the past two years.*

I also encourage couples to do another kind of paired listening. Each of our infertility groups has a Sunday session that husbands are invited to attend. One of the highlights of the day is a paired-listening session in which every couple goes off to a quiet space and talks for six minutes per person. They each spend two minutes on three different topics: something you like about your spouse that you never told your spouse, something you like about yourself that you never told your spouse, and something you like about your relationship that you never told your spouse. Think about it—when was the last time you and your spouse talked about good things for a whole twelve minutes?

Let me tell you, when I say the couples come floating out of that session, I'm not kidding. It's very powerful. These couples start out being crazy about each other, but infertility covers that up with layers of anger, guilt, anxiety, and resentment. This exercise will recapture those great feelings and remind you of what you love so much about your spouse—his creativity, his sense of fun, his devotion to a cause you feel deeply about. A lot of couples start trying to get pregnant right after they marry, and infertility is the first really difficult and challenging thing to happen to them as a couple. And because each responds so differently, because it puts them in two totally different emotional camps, infertility really shakes their trust in the relationship. To do paired listening, focusing on the positive, can prove that the foundation of love and affection and admiration is still there. It reminds you that there are things you love about each other that haven't been tarnished by infertility, as Mary, who has endured five unsuccessful IUI treatments, discovered.

> *Paired listening was absolutely pivotal in the turnaround in our relationship. We still do it every couple of weeks. Life gets so harried. You need time to just sit in a space and have individual attention with each other and really say what's on your mind. Bob and I share a real closeness now. We feel a lot more positive about going into this next cycle. And if it doesn't work, it'll be okay. We're doing it together—it's not just me.*

News and Goods

"News and goods" is a wonderful way to bring some positive thinking into your day and your interactions with your spouse. If you look at the events of any given day, about 70 percent are mood neutral, about 15 percent are negative, and about 15 percent are positive. But when you come home at the end of the day and your spouse says, "How was your day?" what are you going to focus on? The negative, if you're like most people. Most of us focus on the glass as half empty, but news and goods forces you to look at things with a different perspective.

News and goods simply means this: When you greet your spouse at the end of the day, the first thing out of your mouth should be positive. What good news did you receive today? What good things happened? In our mind/body infertility program, we tell our patients that for one week, every day when they come home from work or when their spouse comes home, the first thing they have to do is to ask each other what new and good thing happened to them. This forces you to focus on that good 15 percent of your day. And it forces you to look at your day differently—you have to think about what good thing happened to you so you can report it to your spouse. It forces you to look at your day until you find something good. It doesn't mean you won't talk about the bad things that happened—there'll always be time later in the evening to talk about the negative stuff. Although you may find, as many of my patients have, that once you start focusing on the good parts of the day, the negative parts start losing their importance.

My patients tell me that their husbands love news and goods because their husbands are relieved to have their wives talk about something besides infertility. (Since most aspects of infertility are negative, it compels the first interaction with the spouse to be about something other than infertility.)

Listening to someone else's good news (unless it's a pregnancy announcement!) can make you feel great, as the women in our mind/body infertility groups discover firsthand. The week after I teach news and goods, I ask each participant to describe something new and good that happened the previous week. After each of us talks about what we've done—bought a new sweater, made a reservation for a weekend at a bed-and-breakfast, had a great phone call with a friend—everyone in the room is smiling, because it's fun to hear about these good things. I ask people to imagine how we'd look if we'd talked about the bad things that happened this week. Hearing negative news really brings you down, yet we hear bad news all the time. In our society most of us were brought up to believe that to share good things is bragging but to share bad things is socially acceptable. When you were in school, if you got an A-plus, you probably didn't go around the room telling everyone about it. But if you got a B-minus that you thought was unfair, you'd go bitching and moaning to all your friends. Most of us have been programmed not to talk about the good stuff, but it's OK to talk about the bad stuff. We underestimate how much people do like hearing good news. Of course, you have to watch what you talk about. People don't want to hear that you got a twelve-thousand-dollar-a-year raise and are going to buy a beach house if they're struggling to make ends meet. But they do want to hear about other good things, like the fact that you saw a spectacular sunset, or that a neighbor stopped by with a jar of homemade jam.

Karen, who's had a handful of unsuccessful IUIs and two IVFs, finds news and goods really helpful.

> *Usually, my husband's not home until seven-thirty or eight. I'm usually making dinner when he walks in. Instead of complaining, I focus on the good things. That way we start on a positive, upbeat note. That sets the tone for the evening. It can be easy to fall into the trap of talking about nothing but infertility. It's exhausting to talk about it all the time. Sometimes it just drains energy out of me. So we try to avoid talking about it too much.*

Patty and Don hardly see each other—she works days, and he works nights and occasional weekends.

> *We kind of exist on phone calls, e-mails, and notes. We had always left each other notes, but they usually revolved around our dogs. Now the notes are uplifting, a cheerleader type of thing. What I like to do now is leave him a very cheerful note when I go to bed, so when he comes home at night he's got a happy note.*

There are lots of things that happen to us on a daily basis that are really good. When you share them with people, they don't induce jealousy, but rather they remind people that good things do happen. And when you share good news with your spouse at the end of the day, it allows you both to remember that despite your infertility, there are still many good things happening in your life—and your marriage—that can easily go unnoticed.

 On the Other Hand . . .

Although most couples struggling with infertility learn to communicate, compromise, and become more tolerant of each other's reactions, there are some who find themselves in a more difficult situation. People react to crises in very different ways. Since infertility is, for many couples, the first crisis they face together, it may be disturbing to see your partner's reaction as so different from yours. Since this book is primarily directed toward women, I am going to focus on what to do if his behavior disturbs you.

I always like to assume that the husband in an infertile couple is basically a good guy who is bewildered by his wife's obsession over having a baby, but who, because he loves her, is willing to do what she needs him to do. However, I have worked with too many couples in which this is not the case. I've seen husbands forbid their wives to see a doctor, refuse to provide a semen sample, refuse to allow their wives to proceed with medical treatment, and even cheat on their wives during infertility. I had one patient whose husband would tell her after every unsuccessful treatment cycle that he was going to go out, knock up some babe, bring home the baby, and make her raise it. He never did it, but the threat was awful. Despite lengthy therapy with her (he refused to see me or any other therapist), she stayed with him, and the two of them ended up adopting. I haven't heard from her in years, but the family situation still worries me.

If you are in a situation in which you feel totally unsupported or even threatened by your husband during infertility, you have some options. If he's willing, the two of you need to see a couples counselor who has experience in infertility. If he's not willing, see someone by yourself, so you can decide whether his behavior allows you to feel comfortable pursuing having a baby with him. Sometimes what you learn during therapy might surprise you.

I had a patient, Emily, whose husband, Greg, would not allow her to have infertility treatment. They had been trying to conceive spontaneously for several years, and their physician recommended IVF because of Greg's low sperm count. But Greg wouldn't agree to treatment. After I met with Emily alone for several sessions, Greg finally agreed to come in and see me. Although I had been prepared not to like him (what kind of man would stop his wife from pursuing treatment meant to compensate for *his* physical problem?), he was actually a nice guy. In a few sessions the reason for his refusal became obvious. He had come from a poor, uneducated family and had managed to get himself through college, graduate school, and into a lucrative and successful career in business. He told me that he had learned over the years that the way to overcome obstacles was not to rely on others but instead just to work harder. For him, moving on to IVF was admitting that he was a failure at getting his wife pregnant, and he could not tolerate failure in himself. Following several discussions on the meaning of success and failure, and once he'd learned that sperm production is independent of character, effort, and independence, he agreed that IVF was their best option, and they proceeded with treatment.

Another couple, Chris and Jeff, had exhausted all available medical treatment, including egg donation (Chris was beyond the age cutoff for Massachusetts egg-donor clinics), and Chris was eager to move on to adoption. Jeff, however, was adamantly opposed to adoption. Chris threatened to leave him in order to adopt as a single parent, whereupon Jeff finally agreed to see an adoption counselor. The counselor helped him to acknowledge his grief at not being able to produce a biological child, encouraged him to explore his biases about adoption, and allowed him to come to the decision that he wanted to pursue adoption with his wife. Once he made this decision, he was the one to buy baby-name books, and he chose the name for their adoptive daughter.

So, yes, in some cases husbands are willing to discuss their issues and work to resolve differences between them and their wives. Unfortunately, however, that doesn't always happen. (Keep in mind that since in most of these cases the husbands don't come to see me, I only know their wives' side of the story.) I have seen husbands sabotage their wives' ability to come in to see me; not give their wives messages when their doctors called; refuse to move a work meeting scheduled for the day of their IUI, forcing it to be canceled because there were no sperm to inseminate with; decline to sign treatment consent forms; and simply refuse to acknowledge the infertility at all.

Sadly, I have seen a number of marriages break apart during infertility. Although the majority of couples find over time that infertility makes their marriage stronger, because it gives them the knowledge and the confidence that they can weather a crisis together, for some couples it seems to exacerbate weaknesses that already exist in the relationship.

If you are wondering whether your relationship is going to survive infertility, you are definitely not alone. Do what you need to do to protect yourself: Seek out a trained mental-health specialist, join a Resolve group, or ask your infertility specialist for a referral to someone experienced in the marital impact of infertility. And please remember something important—although ironic: If your marriage feels unstable and you suspect that infertility is not the sole cause of this instability, a baby will probably make the situation worse, rather than better. I have seen several couples who continued with infertility treatment despite severe marital issues and who conceived, but in each case they separated before the babies were even walking. I have even counseled a couple of patients who divorced after realizing that their issues were bigger than their infertility, met new loves, married, and returned to infertility treatment, knowing that any baby resulting from treatment would be welcomed into a loving and stable relationship. All babies deserve that security. Make sure that the two of you are a solid couple before you expand your family.

CHAPTER FIVE

Turning to Family and Friends for Support

In a perfect world your family would be your biggest source of infertility support. Your mother would listen lovingly and nonjudgmentally when you rant about your inability to conceive. Your father would remind your family-planning relatives, before each holiday gathering, not to announce pregnancies at the dinner table. Your siblings would learn all about the infertility alphabet-soup world of IVF and IUI and GIFT and ZIFT that you live in, and would always know, thanks to some kind of magical siblings-only telepathy, exactly what would make you feel better (truffles? a Johnny Depp movie? pizza and a game of backgammon?) after a pregnancy test fails. And your mother-in-law would never, ever, *ever* remind you that you're not getting any younger.

But let's face it: This isn't a perfect world. If it were, you wouldn't be infertile in the first place. Some families are great about infertility. But others just pour fuel on the fire, adding stress, triggering anger, and making infertile couples feel isolated.

Infertility can activate feelings of sibling rivalry that you haven't felt since your sandbox days. It can be especially difficult when a younger sister gets pregnant before you do or produces the first grandchild for your parents. Even if you're the closest of sisters, you may feel irked that she is getting all the attention and just seems to have good luck in her life, and you may feel frustrated by your parents' excitement over the baby. It's especially tough when your siblings are insensitive. One of my patients, Amy, had experienced a second-trimester miscarriage and then couldn't get pregnant again. Her younger sister, Cheryl, got married about a year later and immediately got pregnant. Cheryl began sending weekly e-mail updates to the family describing every detail of her pregnancy, including pictures from each ultrasound. Amy got up the courage to e-mail Cheryl, telling her that although Amy was happy for Cheryl and her husband, hearing every detail of the pregnancy was just too painful for her. She asked Cheryl to please remove Amy from her e-mail list. Well, Cheryl ignored the e-mail and continued sending weekly updates. Amy appealed to their mother, asking her to talk to Cheryl, but their mother was appalled that Amy wasn't ecstatic over the upcoming birth. Amy came to her mind/body group looking for confirmation that she was in the right (which she got instantly!) and ideas on how to handle her family. She decided to ask her husband to delete Cheryl's e-mails before she saw them, and she agreed that her family was so excited about the baby that they were oblivious to her pain. Although it was difficult, she decided she needed to limit her contact with her family temporarily to protect herself, and she sought out support from friends, group members, and one of her husband's cousins.

I have also seen family members be amazingly supportive. I've seen sisters, cousins, and nieces who offer to be egg donors or surrogates, fathers who take their grown daughters to the hospital for daily ultrasounds, and mothers who cry because they can't stand to see their daughters in so much emotional pain. There are many families who want to help but just don't know what to do or say, and there are families who just don't seem interested in trying.

It can be a gut-wrenching exercise, but you need to take a step back and look honestly at the members of your family to see who can and cannot be supportive. Happy surprises do take place—you might find that a parent or sibling who says horrible, painful things during infertility will turn into your main support system during adoption proceedings. You might also find that the crisis of infertility opens your eyes to the true selfish nature of some family members.

Sometimes you need to take a different approach and look at things from their perspective. I had a patient, Joan, who was furious at her mother-in-law for being so intrusive and insensitive about infertility. The woman was constantly asking for updates and every month would say, "Any news yet?" During a cognitive-restructuring exercise, Joan talked about the strain this put on her marriage. "What's your mother-in-law like?" I asked Joan. "What kind of person is she? How does she handle crises in her own life? What's her relationship with her other children?" Joan said her mother-in-law was in fact overly involved in the lives of all her children. She had been pregnant with Joan's husband, her fifth child, when her husband died. She managed to get a job, hold the family together, and single-handedly raise all five children through sheer perseverance. I asked Joan if maybe her mother-in-law had survived by controlling every aspect of her kids' lives. Maybe this coping technique had worked for her during their childhood and was her norm now, and Joan's husband was so used to it that he didn't even notice it. Joan said, "I had never thought of things from my mother-in-law's perspective, but in fact this makes a lot of sense." Joan thought this over during the coming weeks and observed her mother-in-law at the next family gathering. She realized that understanding where her mother-in-law was coming from made it much easier not to be hurt by her comments or behavior. ("Oh, there she goes again trying to gain a sense of control to help herself feel better. This is not my issue, it's hers.") A year later Joan and her husband adopted and found his mother to be their biggest source of support.

Many families have never before experienced infertility and have no idea how it is treated or how they can help someone who faces it. If this is the case with your family, you need to figure out how to talk to your loved ones, just as you must with your friends. Unfortunately, though, families sometimes have issues that complicate communication about infertility, for some of these reasons:

- Infertility is much more prevalent today.
- Most women going through infertility today do not have a

mother who went through it. Thirty or forty years ago infertile women had few medical options; most just adopted. As a result, few infertile women have mothers with firsthand experience of the situation.

- If your mother knows nothing about infertility (and doesn't bother to read up on the subject), her ignorance coupled with her desperation for a grandchild may lead her to say hurtful things to you. Even if she says them unintentionally, they still cause you pain. Or she may smother you with attention, calling you constantly, insisting on accompanying you to medical appointments, and criticizing your choices.
- If your infertility is a result of your mother's having taken the drug DES when she was pregnant, you may harbor feelings of blame, and she may feel guilty. (A therapist can help to sort out these feelings, if they are intense and are causing major problems for you and/or your mother.)
- Many women feel uncomfortable talking about infertility with their fathers, because sex and anything related to it is a taboo subject between them. (And don't even *think* about talking with dads about the process of collecting sperm for a test or treatment.)
- Religious beliefs and conservative ideologies can stand in the way of family support. Catholic families may not endorse assisted-reproductive techniques, for instance. Fundamentalist Christian families may believe that infertility is part of God's plan, and doctors shouldn't toy with God's plan. Orthodox Jewish families may object to the act of masturbating for sperm collection. And if you're a single woman or lesbian, your family may condemn the whole idea of your getting pregnant in the first place and may feel that infertility is your punishment for going against tradition.
- Siblings may be so focused on their own baby making that they are completely insensitive to the fact that you can't get pregnant. Not only do they fail to support you, but they may, because of either an overdeveloped competitive streak or just plain cluelessness, flaunt their fertility and behave poorly.

- Siblings that have supported you through other difficult times in your life may now become, quite mysteriously, emotionally unavailable to you. This may be due to factors you don't know about: Perhaps they had abortions or became pregnant by accident, or maybe they're not happy about being a parent and feel guilty because they don't like being with their children while you're so anxious to have children of your own.
- Siblings may fear infertility will afflict them, too.

Some of my patients are astonished by their siblings' behavior. Kim, for example, who is trying to conceive with donor sperm, was disappointed when her sister, with whom she'd previously been close, failed to take her infertility seriously.

> *My sister wanted me to come with her to a bunch of holiday parties right after my ninth IUI failed and I was accepting the fact that I wasn't just unlucky, I was infertile. I said I didn't want to come, but she wouldn't take no for an answer. I finally said to her, "Don't you realize that I'm dealing with a life crisis unlike any I've ever been through? I'm just barely surviving this. I'm just barely getting myself to work in the morning. I just can't forget it all and go to parties."*
>
> *It's also frustrating how little interest she has in learning more about infertility. She hasn't educated herself at all, and she asks me the most ridiculous questions. If I had a sister who said she was infertile, I'd get on the Internet and learn a few things about infertility so I could be supportive. If she told me she had cancer, I would do those things for her.*
>
> *She also says idiotic things to me that imply that I'm somehow faking infertility to get attention or something. Once she said, "I don't think you're infertile, I think you're just impatient." I was so hurt by that! It's as if she were comparing having a baby to looking for a job, and how discouraged I would get if I'd applied for twenty jobs and got only one interview. I think she thinks I'm imagining my infertility, even though I'm under the care of excellent doctors and nurses. I just don't get it, honestly."*

So how do you navigate your family through infertility? First, you have to figure out whether you're even going to tell them about it. Deciding whether or not to tell your family about your infertility is a very personal decision, and I would never advise you that it is better to tell or not to tell. But dealing with your family is far more difficult if you choose not to tell them. About half of my patients don't tell their families about infertility, either because of some of the potential problems I listed above or because they just don't want to burden them. All I can say is that if your family doesn't know what you're going through, they can't help you and support you and be sensitive to your needs.

Terri learned this when she chose to keep her infertility secret.

> *We didn't want to tell our parents or anyone else what we were going through. I suppose we wanted to surprise them with a pregnancy. But it became very difficult. Finally my therapist told me to treat infertility like any other chronic medical problem. She gave us permission to tell people about it and to tell everyone what kind of support we needed. And that made a huge difference—our families and friends really came through for us after that.*

What if you tell your family about your infertility and they don't come through for you in supportive ways? How do you cope with their insensitivity? How do you manage them in such a way that instead of causing you distress, they can actually provide support? Start by remembering these four words: Educate, communicate, attach, and dodge.

Educate. If your family knows nothing about infertility, teach them. Lend them books, tear out magazine articles, and point them in the direction of helpful Web sites. The more they know, the more helpful they can be.

Communicate. Tell your family what you need. If your mom calls too much because she's desperate to be a grandmother, tell her to stop calling so often. If she doesn't call enough, tell her to call more. If your sister blabs incessantly about her six-month-old, tell her to blab less. If you'd like your mother to cook you and your husband dinner every night for a week after your IVF, ask her to. Unless you let your family know what helps you and what harms you, you can't blame them for all the dumb things they do or say. You cannot expect them to read your mind.

Involve your husband in communicating with both families. Talk with him about what you will and will not tell your families, and then make a plan for who will tell what to whom when. Will you call your mother each time you have news about a treatment or test and have her spread the news to your whole family? Will you call your in-laws, or will your husband? Or perhaps it would make more sense to send both families a letter once a month with an update of the latest test results and conversations with doctors. Will you tell your families everything, or is anything off-limits? By making these decisions now, you can avoid conflict with your husband and family members later.

Attach. If certain family members are being absolutely wonderful, connect yourselves to them. A family member who can listen to you without making judgments can be a real gift. Understand, however, that the person who can fill this role beautifully may not be who you'd expect her to be. It may not be the sister you're so close to—instead, it may be a childless aunt (oh, so *that's* why she never had children) or even a distant cousin with whom you've never before shared secrets or had anything in common (oh, so *that's* why she had triplets). Climb around in your family tree long enough and you might just come across someone who understands your predicament a lot better than you would have thought. Keep in mind, though, that if you suddenly get very chummy with an aunt, your mother may feel jealous.

Dodge. After you've educated and communicated, and certain family members still don't get it, stay away from them. It may be painful (or, who knows, maybe it's a relief) to avoid your sister or your father-in-law, but you've got to take care of your emotional health right now. If there's no one in the family who can give support, go to your friends. Sometimes family members cannot, for whatever reason, give you the support you need. They may not have the capacity or skills to rally around you. If that's the case, recognize your family's limitations and move on. Spend your energy looking for someone who can help you, not complaining about your family's shortcomings.

You may also have to avoid confiding in family members who can't keep a secret. If you don't want anyone to know what you're going through, and your mother can't keep a secret to save her life, you have to make a choice: Tell her and risk having everyone know, or protect your privacy by keeping Mom in the dark.

In-laws are often a whole other kettle of fish. They can be every bit as difficult, insensitive, judgmental, and clueless as blood relatives. But since they're not *your* blood relatives, you may find it even harder to talk with them than with your own family. Since you haven't grown up with them, you probably have very little experience dealing with them.

From what my patients tell me, mothers-in-law are often the most difficult family members during infertility. Perhaps your mother-in-law blames you for the infertility even if it's actually caused by her son's male factor. Or maybe she never liked you anyway, and infertility is just another reason to be against you. On the other hand, some of my patients have found strong allies in their mothers-in-law.

If your mother-in-law drives you crazy with her insensitive comments, and you can't dodge her completely, brush up on your cognitive-restructuring skills. As soon as that searing comment comes out of her mouth, put it to the cognitive-restructuring quick test: Is your mother-in-law's comment logical? Is it true? Say your mother-in-law throws this zinger out at you. "You know, if you didn't work so much, you'd get pregnant." Think fast, girl: Is that thought logical and true? Of course not. Your doctor knows a lot more about infertility than your mother-in-law does, and he hasn't suggested that you scale back your work. If he thought it was an issue, he would have brought it up. Just to be sure, you can check with him at the next appointment. There. Negative thought defused (although the anger you feel from her insensitivity may remain for a while).

Support from Friends

When you're going through infertility, you need supportive friends more than ever. Unfortunately, just as with members of your family, some people don't know how to comfort their infertile friends. Many fertile women have never had an infertile friend or family member, so they are oblivious to what might help or hurt you. They don't realize, for example, that calling you with the exciting (to them) news that they're pregnant could devastate you. They're happy about their news, and they just assume you'll be happy, too—the possibility that their announcement might upset you would flabbergast them, and it's probably the furthest thing from their mind. After all, you've cheered their other good news over the years—their promotions at work, their engagements, their new houses. So if they get pregnant, you should be happy for them. Naturally! Why wouldn't you be elated? And in a way you are happy—you love your friends and you want their dreams to come true. You wouldn't want to inflict infertility on them. But every time someone else gets pregnant, it's a massive reminder that you're not. Pregnancy is a whole different ball game—but even though that is exquisitely clear to you, most likely many of your friends are clueless.

It's tempting to isolate yourself from your friends, to avoid their calls and their parties and their visits. But when you cut yourself off from their thoughtless behavior, you also cut yourself off from all the potential love and support they could be giving you. Chances are your friends *want* to support you through this difficult time, they just don't know how. So your challenge is to teach them how to be a friend to you right now.

Start by figuring out what you need from your friends. Someone to listen to you complain once in a while? Someone you can cry to when yet another friend announces her pregnancy? Someone to sit with you while you give yourself shots or to drive you to doctors' appointments? Someone who can be in charge of e-mailing all your other friends when a procedure fails?

After you decide what you need, think about which friends seem most apt to be able to give it to you. Think about this carefully, because your best allies during infertility may not necessarily be the buddies who helped you plan your wedding or find a new job. Friends who are just starting a family may be too preoccupied to be able to help. Consider friends with older children, single friends, gay friends, friends who are childless by choice, and opposite-sex friends. Or cast a wide net and tell everyone, and see who seems to respond in the most supportive way. You may have to try a few approaches to ferret out the friends who will be best for this crisis, but it's worth the effort, because having a good friend to help you through infertility can be a priceless gift. Even if you have a fantastic husband who is there for you all the time, it's great to have a friend or two who can lend a hand as well.

Whenever you tell a friend about your infertility, make sure she knows what you need. Unless she's had another close friend with infertility, she probably doesn't realize that it drives you crazy when people ask you, the day after your period is due, whether you got it or not. It's your job to let her know how to interact with you. If you want her to ask about your medical procedures, tell her. If you don't want her to show you pictures of her nieces and nephews, tell her. If you dislike it when she shares articles on infertility that she comes across in magazines, let her know. You might even want to make a list of your needs, either just as an exercise for yourself or as an open letter you can hand out to friends. The one thing you can't do is assume that your friends can guess what you want, even if they know other infertile women. The only way friends can know what you need is if you tell them.

Of course, all this presupposes that you have chosen not to keep your infertility a secret. Many of my patients, for a variety of reasons, don't tell friends about their situation. Keeping infertility a secret is the right choice for some people; others are more comfortable spreading the news. Some couples are extremely ill at ease with the idea of discussing intercourse, sexuality, and their reproductive health with others. Plus, they don't want to have to call twenty people every time a period arrives or an IVF fails or a miscarriage occurs. Others have no problem with such openness and feel that the more people they tell, the more support they're likely to receive. You have to balance what's best for you and your husband in the long run. If you're reluctant to broadcast your situation, you can tell some friends and not others—maybe tell your home friends but not your work friends. Or tell family and a few selected friends, but not everybody.

It's tougher dealing with your work friends. A lot of people with infertility tell family and friends but don't want anyone they work with to know they're trying to conceive, either because they fear repercussions or discrimination or because they just feel that it's nobody's damn business. If you don't want your boss to know but would like to tell your good friends at work, swear them to secrecy but keep in mind that some people are terrible at keeping secrets. And remember that even if you don't share your situation with co-workers, they're likely to be suspicious that something is going on, particularly if you're undergoing infertility treatments and missing work or coming in late. Even if they don't know exactly what you're up to, they'll probably figure out that something is happening.

Whatever you decide, remember, if someone does not know about your infertility, you can't expect her to be sensitive to you. Friends can't be supportive if they don't know what you're going through, and they can't help effectively if you don't tell them what you need.

Selective Avoidance

Now that I've tried to convince you of how important it is that you stay connected to your friends, I'm going to go in the exact opposite direction and tell you about selective avoidance. I know, it sounds like the name of an infertility treatment, but it's not. Selective avoidance means staying away from the friends, family members, events, and situations that cause you intense pain. Yes, you need friends, you need family, and you need to go to an occasional party. Yes, you need to educate your friends and family on how they can support you through your infertility. But you don't need to be around insensitive people who remain insensitive even after you tell them what you need. And you don't need to force yourself to attend events that make you cry.

Consider this an official prescription: From now on, you don't have to go to baby showers. There, don't you feel better already? If you like that, I've got more. Here are some other selective-avoidance strategies, and if anyone questions you on them, tell them that these are doctors' orders.

- **Don't push yourself to socialize with friends or family members who are pregnant or who have new babies.** If being near a pregnant woman, nursing mother, or infant makes you sad or angry, then avoid them. If you think the parents or parents-to-be can accept it, tell them the truth, that it's nothing personal, you're just uncomfortable being in such close proximity to fertility. If you think they'll be upset, make up excuses. You can always "catch a cold" that will keep you away from the baby. There's just no reason to put yourself through such agony. But at the same time . . .

- **Don't automatically cut yourself off.** Some of my patients have actually found it comforting to be very involved in their friends' pregnancies or in the lives of their friends' babies and children. "I find it very helpful to embrace my infant nieces rather than avoid them," says my patient Sandy. "I have a feeling that by avoiding them I would just let my fears grow into a phobia. By embracing them I manage to deal with my fears. I didn't push myself against my will, though—it just feels like a natural way for me to deal with this problem."
- **Don't force yourself to socialize with friends who are insensitive to your infertility.** If, no matter how many times you tell a friend that you're doing everything medically possible to have a baby, she continues to insist that you would get pregnant if you'd just stop trying so hard (or drink green tea or see her acupuncturist), feel free to avoid her. Don't feel guilty; you've done your best to maintain the friendship.
- **Stay out of baby stores.** If you need to send a gift and BabyGap makes you cry, go someplace else. Choose Curious George books at a bookstore, buy a gift online, or send a department-store gift certificate instead. For new parents, cook a meal and have your husband drop it off, or have a take-out dinner delivered by a local restaurant or gourmet shop—they'll probably appreciate that more than yet another receiving blanket anyway. You can be a lovely and generous and considerate friend without hurting yourself.
- **Plan adults-only activities.** Avoid your friends' babies by organizing outings to fancy restaurants or shows while the little ones are cared for by babysitters or grandmothers.
- **Hang out with friends who have older children (and vasectomies).** Most infertile women don't mind being around older kids. What they want is a pregnancy and a baby—not a

runny-nosed six-year-old Pokémon fanatic. Plus, most parents of older children are bored with baby talk and have little interest in obsessing about infants, as new parents frequently do.

- **Explain yourself.** Don't just disappear from your childbearing friends' lives without an explanation—let them know why you need to distance yourselves. If you feel uncomfortable talking to them in person, write a gentle letter. Say something like "We really love you guys, and your friendship means a lot to us. But right now we're going through an incredibly difficult time, and it's hard for us to be around pregnant women or new babies. Please be patient with us while we go through this."
- **Change your routine.** Simple schedule shifts can sometimes take you off the path of a pregnant acquaintance. If the pregnant woman you work with eats lunch in the company cafeteria at noon, for example, avoid her by eating at one o'clock or by eating at your desk and taking a mindful walk outside. And if an office baby shower is planned for Thursday morning, schedule client meetings for that time.
- **Split up.** If you are friends with a couple, she's pregnant, and you want to avoid her, that doesn't mean your husband has to avoid her husband. You can stay home by yourself—or, better yet, go out for a pedicure or an afternoon of shopping with another friend—and the guys can socialize at a ball game or over a few beers. That will give your husband some valuable guy time and a break from infertility without forcing you to spend time with the fertile wife.
- **Just say no.** As wives, daughters, and caregivers, women are conditioned throughout their lives to say yes to our families and friends whenever they are in need. But learning to say no without guilt is crucial to your emotional peace and is particularly important when you're suffering from infertility. There are three ways to turn down requests: a simple no; no, but how about this instead; and no, for this reason. Practice saying no to the situations that cause you stress and sap your

energy. As my colleague, psychologist Ann Webster, says, "Sometimes to say no to someone else is to say yes to ourselves."

- **Remember, this is a temporary crisis.** You're not going to be infertile forever. Either you'll get pregnant or adopt or choose to remain childless. And when that time comes, you can resume your interrupted friendships. This is a temporary crisis, even though it feels as if it will never end.

CHAPTER SIX

Infertility and Your Career

Most women with primary infertility—the inability to have a first child—work outside the home. Some are career women who have spent years scaling the corporate ladder. Others enjoy their jobs but don't have any particular attachment to them. And some women are just marking time in jobs they may or may not care for, waiting to get pregnant. They never expected to have a long career—they'd just planned to work a few years, get married, have children, and then stay home and make a career out of full-time motherhood.

No matter how you feel about your job or what career path you chose, infertility is most likely making a mess of your plans. If you're a high-powered executive, infertility may be hampering your ability to move ahead in your career. If you're an hourly worker, your employer may be growling about your repeated requests for time off for medical appointments. And no matter what you do for a living, you're probably having trouble focusing on your job as the emotional roller coaster of infertility interferes with your concentration.

If you're like most infertile women who work outside the home, you are plagued by some of these worries and concerns:

- As you try to juggle the demands of your career with the inconveniences of infertility, your job performance may be suffering. This is deeply embarrassing for you, because you've always been proud of being a conscientious employee with a strong work ethic.
- As your personal priorities shift, your work may not fulfill you anymore. You may find yourself losing interest in a career that, before infertility, invigorated you.
- If you work long hours or travel a lot for business, you may be concerned that your job is contributing to your infertility.
- You may feel trapped in a job you no longer enjoy, but you're reluctant to look for a new one. You hang on, lacking the energy to job hunt. Besides, if you conceive, a new boss will be annoyed to have a pregnant new employee who'll need time off for maternity leave.
- You might even be so overwhelmed by the time commitment required by high-tech medical treatment that you're thinking of leaving your job to devote yourself full-time to resolving your infertility. But can you afford to quit? And if you can, is it even a good idea?

In this chapter I'll discuss the many problems that women face as they try to balance their jobs with their infertility. I'll also share a variety of solutions that have helped my patients to concentrate better on their jobs and make wise decisions about the questions and choices that arise as they negotiate the competing demands of work and infertility. No solutions are right for everyone—for one woman the right choice is to work less, and for another to work more. But by considering all the options and listening to the stories of other infertile women, you'll be better able to perform in your job and make the decisions that are best for you.

Let me begin by telling you about Martie, a patient who loved her job but, because of her infertility, felt unable to give it her all. She faced many of the problems that infertile women often confront.

Martie's Story

Martie, a forty-two-year-old businesswoman, married at age forty and immediately started trying to get pregnant. Both she and her husband adored children and were anxious to have a family. Nothing happened during several months of trying, however, so Martie sought medical opinions about what she should do next. Tests showed that neither Martie nor her husband had physiological problems that would prevent conception, but her OB/GYN suggested that they start medical interventions immediately anyway, because of her age. "We wanted to be aggressive, because my chances of getting pregnant were decreasing every month," Martie says. She quickly proceeded from ovulation drugs to IUIs, but still she didn't conceive.

At the time Martie worked in a job that required a tremendous amount of her energy, and she couldn't help but wonder if the stress of her job contributed to her infertility.

"I was the chief operating officer at an Internet start-up," Martie explains. "And my husband was working at a start-up company as well. We both worked constantly. When the founders of the company offered me the job—they had approached me about working for them—I told them I couldn't take it because I was trying to get pregnant. And they said, 'Don't worry; we'll just work around it. We want to be a family-friendly company.' They were very supportive, so I took the job."

Things started out well, but soon became hectic. "We would work nights and weekends, because we were trying to get funding. We'd have meetings that wouldn't even start until eight P.M. As time went on, the economy started to slow down and money got tight. I liked the job, but I found that I just couldn't stay committed to the company. In a situation like that, you've got to give 150 percent—anything less than that isn't fair to everyone else in the company. You are so split because you feel that you're supposed to be doing your best and giving it everything, but you're moody because of the drugs, or tired, or you have to go for all these appointments. Even though the founders of the company were still supportive, it just didn't feel right."

So Martie made one of the most difficult decisions of her life: She quit her job. "I love working, and I'd never not had a job. Quitting was huge. But my husband and I were committed to getting pregnant, and I thought, 'I'm working a lot of hours, I'm working nights and weekends, and I'm stressed out by it. If one of the factors in my infertility is stress, we can take that factor out of the equation.' So I quit, and we built our whole life around making everything optimal for me to get pregnant. I needed to know that I had done everything I possibly could to get pregnant."

Being home was tough for Martie, and she struggled to find things to do. "I missed working. I was gaining weight because of the infertility drugs. I tried to exercise, but I was so tired and moody I couldn't. It was really hard for me." To fill her days, Martie did repair work around the house, organized years of photos into albums, and spent time each week caring for a friend's babies. "People told me to just sit around and read books, but that isn't my nature."

After she quit, Martie underwent two IVFs, which failed. She was distressed and tempted to give up. "I was definitely at the end of my rope. I was thinking of adoption, of going back to work. I just couldn't sit around waiting to get pregnant anymore—I couldn't stay in that holding pattern."

With a bit of reluctance, she tried IVF one more time. That time she became pregnant with twins, and her babies are due in a few months.

Eye off the Ball

Even for the most dedicated employees, it's difficult to concentrate on your job when you're in the middle of infertility—for example, when you're on day twenty-seven of your cycle and hoping not to get your period, or when you're waiting for a phone call from your doctor with test results, or when you've got to play hooky from a meeting on day fourteen of your cycle so you can sneak home to have sex with your husband before he leaves town for a weeklong business trip. You may find your mind wandering, your patience slipping, and your interest in your job, your co-workers, and your customers disappearing.

The fact that infertility cuts into your schedule left and right doesn't help either. If you're having medical treatments, you're in the hospital or clinic so much that you spend more time chatting with the receptionists than you do with your best friends. If you're trying to get pregnant naturally, you and your husband both have to be home mid-cycle, and for some couples, that's impossible. Take Trudy, for example.

Trudy, a department-store manager, had little trouble arranging her own schedule to accommodate her infertility, but her husband's schedule was another story. Ted is in the military, and his job required him to be overseas for months at a time. "When we first started trying to get pregnant, nothing happened for three or four years. But I never knew if we really had a problem or if we just weren't lucky with our timing." Eventually Trudy and Ted decided to pursue infertility treatment. Three IUIs failed, and now they're moving on to IVF with Ted's frozen sperm. "We just can't work around his schedule anymore," Trudy says. "Now he doesn't even have to be here—we're going ahead without him." She adds with a laugh, "People around here are going to wonder how I got pregnant if it happens while he's gone!"

What's more, daily or every-other-day ultrasounds, a requirement for IVF and other assisted-reproductive technologies, make you late for work and wreck your morning schedules. With any other medical procedure, you usually know when your appointments are and you can schedule work responsibilities around them. But not infertility treatments—they keep you on your toes. If you're doing a high-tech cycle and you usually ovulate around day fourteen, for example, both you and your husband can plan to be around that day. But you may ovulate a day earlier, or a day later, and when ovulation calls, you both must answer, no matter who has a crucial meeting or a boss who fumes when someone calls in sick. I've had so many patients who've had an incredibly crucial presentation, trip, or client meeting, and their IUI was either earlier or later than they thought it was going to be. They ended up in an enormous work crisis.

And then there are all the little predicaments that add up to big stress. Your doctor is scheduled to call you with test results in the afternoon, but you miss the call because you're with a customer. You call back, but he's busy with another patient, and you end up swapping voice mail until the next day. Or you get the call, but because you have no private place to talk, you can't ask the doctor all the questions you'd like to ask—why can't he just call you at home in the evening! Or the company receptionist screens all the calls, and you hate the idea of her knowing that your doctor is calling *again*.

My patients have come up with several solutions to these problems: Some ask clinics to leave lab results and other news on their home answering machine. That way they can listen to the news in private, and if it's bad, it won't wreck their concentration at work for the rest of the day. Some give the clinic their cell-phone number, and when the phone rings, they dash off to a private office, a conference room, or even the bathroom for seclusion. If a snoopy secretary is a problem, ask your clinic nurse to identify herself as your cousin Lucy or your company's new client from Terre Haute.

It also helps, if it's at all possible, to use smart scheduling and planning as a way to insulate yourself from infertility-related stresses that occur at work. Many patients find simple things that can help them. For example, if you have control over your own schedule, don't plan meetings or anything else important for times that you know will be difficult. If you can swing it, arrange out-of-office appointments, paperwork, or behind-the-scenes backroom work on days that you're best off being away from other people. Try to think ahead about what you need to help you through emotional outbursts. Stock your office space with tissues, eyedrops, under-eye cooling gel (for reducing post-tear puffiness), and a makeup bag. Know which is the most private bathroom, and use it on day twenty-six and beyond, so that if and when you get your period, you can cry without being overheard. Keep a few pieces of chocolate in your desk if that kind of little pick-me-up works for you. Think about what will help you survive difficult moments, and plan ahead for them as much as you can.

Pregnant co-workers add another unpleasant stress at work. If you have trouble being around pregnant women and you have friends and family members who are expecting, at least you can choose to avoid them—you can skip family parties and stay away from social events. But co-workers you're stuck with—you can't avoid a pregnant manager, receptionist, or assistant. One patient of mine had four pregnant co-workers at one time. Going to work became a nightmare for her, because these women didn't know she was infertile, and since she'd been married for a while, they suspected she'd be trying to conceive soon. Unfortunately, they automatically considered her part of their little pregnancy club and talked constantly with her about babies.

It's enough to drive you crazy.

 Work Stress and Infertility

OK, say your job *is* driving you crazy. You're overworked, overwhelmed, and overwrought. Does that mean your job is contributing to your inability to conceive?

I don't believe so. I have not seen any reliable research suggesting that work stress would affect your ability to get pregnant. A European study that was never replicated by researchers elsewhere concluded that women who work outside the home had slightly lower pregnancy rates with IVF than women who didn't work outside the home. But that study didn't control for other variables. For example, did the working women drink a lot more coffee or get a lot less sleep than the ones who stayed home? Did the researchers control for depression? These are important questions that were not addressed in that study.

So, no, I don't believe that job stress would hamper your fertility. If your job is causing you to be depressed, however—either because you're in a job that forces you to be around pregnant women and babies or because you hate your job and would be depressed by it whether or not you were infertile—that concerns me. As I said in Chapter 1, the research indicates that depression can indeed affect fertility.

There are other job-related circumstances that may prevent you from getting pregnant, but those are exceptions. Obviously, if your work requires so much travel that you and your husband barely see each other, let alone make love midcycle, that's an issue. If you're working in a job that exposes you to environmental toxins, or if you do shift work and have to rely on massive amounts of caffeine to stay awake, or if you're so stressed at work that you have to drink a few beers or smoke a pack of cigarettes every night to calm down, those are legitimate worries. But there is no research that I'm aware of to support the idea that just the stress in and of itself will make you less fertile. And I have not seen, in my clinical practice, a rash of successful conceptions and pregnancies among infertile women who stop working outside the home, compared with those who hold jobs.

Should You Tell?

Sometimes telling your boss that you're being treated for infertility can reduce stress—at least she'll know you're sneaking off for ultrasounds and not secret rendezvous with competitors. But letting your boss or co-workers in on your situation isn't always the right choice. Infertile women often struggle with the question of what—if anything—to tell their bosses and co-workers about their situation.

On the one hand, it can be helpful not to have secrets, because then you don't have to keep making up excuses when you miss work. (Monday: "I have to wait for the cable guy." Tuesday: "The cable guy didn't come yesterday, so I have to wait for him again today." Wednesday: "Can you believe it? The cable guy blew me off *again*, but he *swears* he'll make it today.") If you don't tell, suspicions will mount over time as co-workers watch your comings and goings with growing wariness and either figure out what you're doing or devise cockamamie theories about your whereabouts. It doesn't usually take much for co-workers to conclude either that you're facing some sort of illness that you don't want to talk about or you're looking for another job. About ten years ago one of my patients had a co-worker come up to her and say, "You know, I've been noticing that you're out a lot, and I just want to tell you that I'm terribly sorry that you have AIDS. How much time do you have left?"

On the other hand, when you tell, you risk being plagued with questions ("So, are you pregnant yet? No? Why not? What's taking so long?") and missing out on promotions and key assignments. A number of my patients have told their bosses and then stopped getting good assignments. The boss writes them off, thinking they're going to be distracted while trying to conceive, preoccupied for nine months of pregnancy, and then missing in action on maternity leave. Other patients have had good luck. One patient agonized over whether to tell her boss. She finally did, and it turned out the boss was the mother of GIFT twins and was extremely supportive and encouraging.

What's the best solution? My patients ask me that all the time, and there's no right answer. It's a very personal decision that depends on the relationship you have with your boss, what kind of company you work for, whether you will be penalized, and so on. It also depends on who your boss is—if she has twins or became a parent in her late thirties, she might be more open than a sixty-year-old man with grown children or a fifty-year-old woman who never married. I lean slightly toward openness, but I know that for some women, this is absolutely the wrong approach. Fortunately, it worked out well for Andrea, a program director for a health-care facility.

> *I wanted to tell my boss that I was starting infertility treatments, but I was worried. I thought she might think I had less of a commitment to my job, or that I would quit my job when I had a baby. I was worried that she was going to imagine that I'd be thinking about babies all the time and not about my job. But I was forthcoming with her, and she was great. I told her I would be late for work a few mornings a week, and she said it was fine. Then, after I got pregnant, I had to stay in bed for a while, so I ended up working on bed rest for three months. And she supported me the whole time.*

If you can't decide, perhaps a trusted co-worker who knows your work situation well can give you some perspective. Or a few sessions with a career counselor may help you find the right answer.

Keep in mind, however, that even if your co-workers do know about your struggles with infertility, it doesn't mean they'll understand what you're going through or treat you with compassion. Even your closest friends don't always know how to help you—and they love you. Coworkers may know even less about what you need than your friends do. That's why I recommend educating your co-workers about what you require from them. (You may be in a work situation where you can't do this. If not, keep a journal with you for venting, and make sure you have a friend you can talk to in the evenings.) If you don't want them to ask you about your treatments, tell them so. If you like it when they express concern for you, thank them and reinforce this behavior. If you'd prefer that expectant co-workers not discuss their pregnancies in front of you, gently explain this in the same way that you would explain it to a friend: "I'm really happy that you're pregnant, and I wish you well. Ordinarily I would like to hear the news of your pregnancy, but right now it only brings me pain. Would you mind if we don't talk about it very much?"

As with friends, some co-workers will continue to say the wrong thing—they may even say rude or hurtful things—despite your best efforts to educate them. When this happens, and you can do nothing to limit co-workers' boorish behavior, try avoiding them if possible, and when you can't, reach into your box of coping skills for ways to make it easier to put up with them.

Coping at Work

Among the most potent techniques for keeping yourself from getting steamed up at work are mini-relaxations.

I mention minis a lot in this book, because they're good for just about anything that ails you. But they are especially helpful at work, where stress can flare suddenly and where, more than just about anyplace else in your life, reacting in an appropriate, noninflammatory way is crucial. At work you've got to stay in control of your emotions no matter what crops up, and minis can help you do that beautifully.

Unfortunately, infertility can rob you of your capacity to be cool and professional at work. Infertility drugs and bad news can trigger mood swings, irritability, and overreaction everywhere in your life. But let's face it, if an infertility-related mood swing pushes you to lash out at your husband or your mother, you can apologize afterward and they'll probably forgive you. Co-workers, clients, and customers may not be as forgiving, whether or not they know you're infertile. Irritability may annoy your husband, but he'll (I hope) get over it. Irritability at work could cost you your job.

Minis are an excellent antidote to work stress because they can be done anytime, anywhere, and they yield immediate results. They help in two ways: Physiologically, they shift you instantly from shallow chest breathing to deep abdominal breathing, which rushes oxygen to your body and helps stop the fight-or-flight stress response in its tracks. And psychologically, minis give you a chance to step back from your stressor and gather some perspective about how best to react to the situation. Imagine you're in a large meeting and your co-worker has just said something inflammatory to you. Your first reaction is to lash out at this ignoramus, verbally rip her to shreds, strangle her, curse her and the horse she rode in on, and show everyone else in the meeting what a fool she is while you're at it. Infertility can cause some very dramatic—and not always rational—reactions.

It's unnerving, but when you're infertile you can't always trust your instincts. Yes, maybe this co-worker is an ignoramus. But putting her in her place in front of half the company may not be the right strategy. Instead of lashing out, do a mini. Take a deep breath, count slowly from one to four, then exhale, counting slowly from four to one. While you're exhaling, tell yourself something that will give you perspective, something like "Survive this moment" or "Defuse this bomb" or "Don't blow up." If you can possibly manage it, repeat this exercise a couple of times. When you finish, you'll be better prepared to address your colleague calmly, in a way that you won't regret later on. Try to answer her charge in a calm, measured way. If you're too emotional to do this, look for an out. Say something like "I don't think this is the time or place for us to hash over that issue. Let's meet privately later to talk it through." If your tormentor persists, excuse yourself from the room to gather your composure. Sometimes backing down from a fight is a much better solution than engaging in a bloodbath that you'll bitterly regret later on.

Minis help in dozens of other work situations, too. You can do minis when a customer or client annoys you, when you have to sneak in late to a meeting after an IVF ultrasound, when the copy machine breaks down, when your computer freezes up. No matter what the situation, minis allow you to step back, get in control, and plan an appropriate response.

Mini guided imagery can be a great (and fun) coping tool at work, too—especially if you do longer guided-imagery exercises at home. Here's a great tip from Fred L. Miller, a personal and corporate coach and author of the upcoming *How to Calm Down* (Warner Books, 2003). He recommends sneaking off to the bathroom when work stress flares up. When you're about to lose it in a meeting or in front of a customer, excuse yourself and head for the restroom. "Lock the door, sit on the toilet, close your eyes, take three deep breaths, and then take yourself to your favorite place in nature for two minutes," Miller says. "After all, they can't keep you from going to the bathroom."

Longer relaxations can help during the workday as well. While they're probably not a substitute for at-home relaxation sessions, lunchtime relaxations can dissolve tension and calm you down, putting your morning in less-stressful perspective while preparing you for the afternoon ahead. If you can find a quiet place, listen to relaxation tapes or guide yourself through a familiar meditation. If noise is inescapable, take a mindful stroll during your break. Even if you work in the city, focusing on the sights and sounds around you in a mindful way can help take you away from your concerns at work.

Connecting and Communicating

Sharing your thoughts can also help keep you centered when work and infertility wad themselves up into one big ball of stress.

Try talking to yourself via a journal. It's a perfect place to let your emotional flare-ups play out. When something sets you off at work, whether it be job related or infertility related, grab a few moments, reach for your journal, and spew out every bad feeling and negative thought that is gushing through your mind. Say what you want about your idiotic boss, your moronic co-workers, your poorly managed company. String together expletives about your selfish doctor who never calls you when he says he will, or about your thoughtless sister who breaks your work concentration with a poorly timed call to tell you that a cousin is pregnant. Say what you want, release your anger, and then shut the book. And make sure it's hidden in a safe place, where co-workers won't find it.

A journal can also serve as a receptacle for worries. You know how when you're doing a relaxation exercise and a thought or worry enters your mind you're supposed to acknowledge it and then let it go? That's a great idea for work, too. If you're focusing on the Smith case and you suddenly start worrying about whether your sister-in-law might be pregnant—after all, she didn't have wine at dinner Saturday—the best thing to do is acknowledge your worry and then let it go. But really, who can do that every time? Especially if you're working on something that's less than scintillating, it can be very hard to let go of distracting thoughts and worries. That's where a journal comes in handy. When an uninvited concern starts buzzing around in your head and breaking your concentration, try stepping away from your work and taking a few minutes to jot down that thought in your journal. Write out your concern, along with a quick list of emotions, questions, or fears. Then put the book aside and see if your concentration returns. For some women, just getting their issue down on paper can take it out of their head and free them to return to their work.

Communicating your thoughts to another person on the job sometimes makes the workday easier, too. Finding someone to talk to can be hard, especially if the pals you usually hang around with are pregnant, new mothers, or unable for other reasons to be effective confidantes for your infertility woes. My patients often find that sympathetic listeners come from unexpected places—a much older mother of grown children in a different part of the company, a colleague with whom you are at odds over work-related issues, even a male colleague with a history of infertility. You never know. But no matter where it comes from, social support can sustain you through difficult times.

The best friend in a case like this is often a woman who has struggled with infertility at some point in her life, although personal familiarity with infertility isn't necessarily a requirement. You may think there are no such people in your workplace, but that may not be true. Look around—is there someone else who's been married awhile and hasn't had children? Does she disappear mysteriously every now and then? She might be undergoing infertility treatment. Is there someone with twins or triplets? Maybe she conceived through an assisted-reproductive procedure. Is there an older woman who never had children? Perhaps she was infertile at a time when medical science had little to offer her in the way of help with becoming pregnant.

To connect with these potential allies, start a very open-ended conversation. Mention a TV-show episode or a movie in which one of the characters is infertile. Or drop something into the conversation about an infertility-related news event. This may be just the opening your coworker needs to admit her experience with infertility. It may seem like something of a fishing expedition, searching for people who will offer understanding social support at work, but if you find someone, you'll be glad you took the trouble to fish. You'll feel far less isolated, having someone to confide in at work and to unload with after an anxiety-producing event.

Thinking About Work

A while back I had a patient who used cognitive restructuring to get to the bottom of some very negative thoughts she had about her job. Cognitive restructuring is an excellent way to fully understand complicated feelings toward your career. Tara was a very high-level financial expert who routinely worked sixty to eighty hours a week. After many years of infertility she ended up getting pregnant on a high-tech treatment cycle, but at eighteen weeks she miscarried. As she was starting her second round of treatment, she joined one of my mind/body infertility groups.

During our cognitive-restructuring session Tara admitted that she was grappling with this thought: She blamed her miscarriage on herself. "If I had worked less during the first trimester, I wouldn't have miscarried," she declared. I told her there's no research to support the idea that working hard causes miscarriages. And I pointed out that women around the world literally labor in the fields doing work that is far more strenuous than financial work and have perfectly healthy pregnancies. But this information didn't ease Tara's mind.

Just then someone else in the group said, "You want that to be the case, don't you? You want your job to be the cause of your miscarriage." Tara started to cry and finally said yes, that was true. Well, sometimes the therapist is the last to understand these things, so I said, "Clue me in here, guys—why does she want that to be the case?" And a group member said, "Tara wants to blame the miscarriage on her job because if she knew that it was caused by working hard, then if she doesn't work hard next time, she won't miscarry." Ah, I get it. "That's true," Tara admitted. "That's it exactly. If I knew that my work caused the miscarriage, then at least I would have some control over what happened, and I could change my behavior next time." Clearly Tara was looking for something on which to blame her miscarriage, and her job was a convenient scapegoat.

After that exercise Tara felt better about her work. And after her next successful cycle she cut down on her work and delivered a healthy child. In my opinion the pregnancy would have survived whether or not she cut down on work. But by making peace with her feelings about her job, Tara removed a piece of worry and stress from her mind. And by cutting down on work while she was pregnant, she helped to eliminate an "I should have" that could have plagued her in the future, had her pregnancy failed.

This is an important point, and it relates both to work and to other choices you make while you're trying to get or stay pregnant. After you miscarry or have a failed treatment cycle, you have a tendency to look back and say, "I should have done this" or "I shouldn't have done that." Whether or not there's scientific evidence to back up a certain "should have" or "shouldn't have," if you suspect that something you're doing may endanger a pregnancy, you should probably think hard about not doing it. Women have to listen to their own intuition when making important decisions. Not that you should be ruled by intuition—remember, a few pages ago I told you that if your intuition tells you to strangle a co-worker, you might want to step back and give yourself a bit of a reality test. But if you have a nagging feeling that something is right or wrong for your infertility treatment or pregnancy, I suggest that you consider letting that feeling influence your action—within reason, of course.

Here's a story to illustrate my point. A few winters ago a patient of mine named Doreen called me in the middle of an IVF cycle. She'd been invited to go on a ski trip during the third week of her cycle, which would have been about a week after the embryo transfer. Her physician said it was okay for her to go, and she asked me what I thought. "Suppose you fall?" I asked. "Then suppose you find out a week later that the transferred embryos didn't take, and you're not pregnant? You would blame it on that fall." A fall probably wouldn't cause Doreen's embryos not to implant—people go skiing and fall all the time and still get pregnant. But Doreen would probably always wonder if that choice had caused the procedure to fail. She would beat herself up for the rest of her life about it. You don't want to do anything that will cause you to say, "If only I hadn't." That goes for ski weekends, and it goes for work as well.

So if you're having negative feelings about work, put them to the cognitive-restructuring test and examine them for their ability to cause stress, their origin, their logic, and their truth. Restructure them in a more accurate light. Then you'll be better able to make decisions about how to put your infertility and your career into context with each other.

What About Quitting?

Leaving a job—either temporarily, with a leave of absence, or permanently—is an option that some infertile women consider. Undergoing infertility treatments can feel like a full-time job in itself, considering how much time, energy, thought, and planning the treatments require. Giving both a job and the quest to have a baby all the energy they each require is impossible for some women. Quitting used to be relatively unusual, but now I see larger numbers of women in my infertility sessions leaving their jobs to devote themselves to getting pregnant.

The advantage to quitting your job, if you can afford it, is that you can spend all the time and attention you need for finding and pursuing the best infertility treatment, without the distraction of a job. You'll have plenty of time to do relaxation exercises, meditate, self-nurture, and take walks. You won't have to steal off to doctors' appointments or make up excuses about getting to work late. And, of course, you'll be able to lie in a hammock and go for walks in the woods and do all those other things that working people dream of doing instead of schlepping to work every day.

The disadvantage is that if you leave your job, you will be extraordinarily focused on infertility. Instead of being a secretary with infertility or an accountant with infertility, you become a professional infertility patient who thinks of little else other than getting pregnant. Quitting work can also cause money problems, and if your health insurance doesn't cover infertility treatments, quitting may be financially impossible. If quitting puts an enormous financial burden on you and your husband, not working might be more stressful than working.

Problems may also arise if your job figures heavily into your perception of who you are. If you think of yourself as Elizabeth the librarian or Elizabeth the baker and suddenly you're just plain old Elizabeth, will you be leaving behind too much of your identity? Will you feel whole being just Elizabeth? For some women, the answer is a resounding yes. But others would feel lost, especially if all they want in the world is to be Elizabeth the mother. If much of your identity is based on your career and you leave your job, then who are you?

A leave of absence may be a good compromise. A leave that coincides with surgery or an IVF cycle gives you time to recuperate, rest, and plan your next step, should this step fail. It also gives you more time to do relaxation exercises, take yoga classes, and attend support-group meetings, which may aid in healing or treatment success. In my opinion, taking a month off when you do an IVF is a great strategy if you can possibly manage it.

Or you can set a longer time limit—say, a year. A number of my patients have taken a year and immersed themselves in infertility treatment. They do three or four intense treatment cycles, they devote themselves to relaxation and other mind/body strategies, and they sustain lifestyle habits that are the very best they can be. If they get pregnant, great. If they don't, at least they can look back at the year and say, "I did everything possible to get pregnant," before moving on to donor eggs or adoption or choosing childlessness. They are far less likely then to say, what if?

Helen had endured six years of infertility. During that time she'd tried injectible drugs, IVFs, and GIFT, and had suffered six miscarriages. At the age of forty-four, she decided to try one last time to get pregnant using donor eggs. Helen, a florist, decided to take a leave of absence from her job. She signed up for my mind/body class and put every bit of her energy into getting pregnant. "I did every single exercise with my whole heart. My husband and I decided we were going to go for this like getting a Ph.D. We were going to put all our resources and time and energy into this, and if it didn't work, it didn't work." Luckily for Helen, it did work, and she conceived and gave birth to twin girls.

Keep in mind, however, that taking time off isn't for everyone, as my patient Eileen, a stockbroker who couldn't conceive because of high FSH levels, proved:

When I was in Dr. Domar's mind/body program, I was really, really focused on my job. In fact, I was so focused on my work and I worked so much that some of the other women in the class said, "Why aren't you concentrating on what you really want? Why don't you quit your job?" We talked about it, and eventually I realized that, for me, work was a safe haven. I could come in to work and just throw myself into it and forget about infertility.

If you do leave your job for an extended period of time, I have some advice for you. Make sure you've got other things besides infertility to occupy your mind. I worry about someone who quits her job and does nothing else at all but try to get pregnant. Your whole week is focused on that Thursday three-thirty doctor's appointment, and it's easy to just rattle around the house, eating too much, researching infertility on the Internet, and obsessing about whether you'll ovulate Wednesday night or Thursday morning. Do volunteer work, tackle projects around the house, or take up some new hobbies—just be sure you don't sit around. And be aware that if the sight of mothers and babies upset you, there's no escaping them during nine-to-five work hours. If you're out and about during the day, you'll see mothers and babies everywhere you look.

Changing jobs or quitting may be your only option if your job forces you to be with babies or children. Some infertile women are okay with this, but for some, going to work every day causes pain and depression. I've had patients who are obstetricians, labor and delivery nurses, nurse midwives, nursery-school teachers, pediatricians, and even infertility specialists. These are people whose lives revolve around pregnancy, babies, and children. Many of them chose these careers because of their intense love for babies and children. But infertility changes that. Infertile obstetricians and nurse midwives can't stand listening to pregnant women complain all the time. Day-care and preschool teachers' hearts ache when parents drop off their children for ten hours a day. One patient of mine owned a day-care center and had to sell it. It killed her to see parents spending so little time with their children. "If I had my own baby," she told me, "they'd have to pry it out of my arms with a crowbar."

Jenny, an infertile obstetrics nurse, had a terrible time dealing with her pregnant patients.

It wasn't delivering the babies that was so bad, it was the prenatal care that was the hardest part. During the prenatal visits the women complained about every little ache and pain of pregnancy. It was really hard to listen to these complaints. Pregnant women can be very self-centered, and that's fine and normal, but when you can't be part of it, it's pretty hard. I had one patient who was shopping around for practices say that all obstetrics nurses should have children, or they can't be good nurses. Well, two of us don't have children because of infertility. I was ready to cry or wring her neck—I didn't know what to do because I was so mad, but I had to be professional. Looking back, I should have said, "Some people want to have kids but can't." But I didn't say anything.

After several years Jenny conceived via IVF.

> *If I hadn't gotten pregnant, I couldn't have continued to be an obstetrics nurse. It was too painful. The work is hard, and you have to really feel passionate about it to be able to stay up all night, work nights and weekends, and have such a crazy life. It requires giving of yourself to people, and I was mad at those people. I was so envious of what they had. It's terrible to say that, but it's true. My life revolved around pregnancy and birth—that is what I'm interested in, it's how I spent the last fifteen years of my professional life. To be cut out of that club was pretty devastating. I felt like my career was on the line as well as my family-building dreams.*

Many of my patients have found solace in volunteer work. It can provide a welcome distraction from infertility. Occasionally it can also bring fulfillment and joy. Sally, a thirty-three-year-old patient who has been unable to conceive for two years, became a Big Sister to an eight-year-old girl, which brought her enormous fulfillment.

> *I had so much maternal energy that was going to waste. Finally I thought, "Why don't I use some of that on a child who needs it?" So I became a Big Sister. It's not always easy—she's a little difficult at times, and she's not always the easiest kid to love. But the rewards I get from being with her and working with her are amazing. She's not my child, and she's not an ideal child, but I have a lot of affection for her. It's also proven to me that it's very, very possible for me to love a child that is not my own, and that I'd be fine with the idea of adopting a child if I don't conceive.*

I've also had patients who have found great joy in volunteering at animal shelters—one even started a dog-walking business. In fact, an enormous number of my patients have gotten puppies. Dogs give unconditional love, which can be very healing for a woman struggling with infertility. A dog is certainly not a substitute for a baby, but if it can help you through the difficulties of infertility, then so be it. Just remember that if you do adopt a puppy, you may find yourself with a newborn baby and a rambunctious year-old dog. If you don't think you can handle a newborn and a dog, opt to volunteer at a shelter. Or offer to take care of your neighbor's dog once in a while.

If infertility necessitates changes in your career plans, try to remember that the changes don't have to be permanent. You can resume your career climb after you have a baby, adopt, or decide to remain childless. Even though infertility seems infinite now, believe me, it will end, and your life and career will return to normal.

CHAPTER SEVEN

Special Cases: Secondary Infertility and Infertility in the Unmarried Woman

Infertility doesn't strike just married couples who are trying to conceive their first child. A woman who is struggling with infertility may also be single, or a lesbian in a committed relationship, or a married woman who has a first child and can't conceive a second, which is known as secondary infertility. Although the people facing unmarried or secondary infertility must cope with many of the same problems as a married couple trying to conceive for the first time, they also face additional issues. In this chapter I'll examine some of those issues and suggest mind/body strategies that will make coping with these challenges easier.

Secondary Infertility

Women with secondary infertility are the Rodney Dangerfields of infertility—they get no respect. Other infertile women can't stand them. After all, they've got one child—isn't that enough? Family members don't understand them. They got pregnant once, why can't they do it again? And when a woman with secondary infertility brings child number one along on a visit to the IVF clinic, others in the waiting room stare and throw emotional darts at mother and child. Some waiting rooms won't even allow in anyone under the age of eighteen because of the bad feelings it can cause.

Statistically, secondary infertility—the inability to conceive and deliver a second child—is actually more common than primary infertility. A woman often conceives easily the first time, and then a couple of years later, when she tries to have a second child, she can't. Or a woman who went through infertility the first time finds that the procedures that succeeded in the past are failing the second time around. For women going through a second bout of infertility, just the thought of starting treatment again brings back all the pain, frustration, and agony. I've had patients who, despite their extreme desire to have a second child, don't pursue treatment because they can't bear to face a second trip into the world of infertility.

It's such a difficult situation to be in. You desperately want a second child, a sibling for your first child, a family for yourself and your husband. But you can't conceive, and you're overwhelmed with frustration. It worked the first time. Why won't it work the second time? This feeling can be especially strong among women who are infertile for the second time; that is, they had primary infertility, conceived with treatment, and now have secondary infertility. They know that with treatment they are capable of conceiving and gestating a healthy baby, but this time it's not happening. Why not?

That's what your friends and family may think, too. They don't understand secondary infertility. They probably think that either you can have children or you can't, and if you had a first one, you can have a second one. Friends and family may not take secondary infertility seriously. If you'd just calm down and pay a little more attention to the calendar, they may say, you'd get pregnant, no problem.

In addition to being misunderstood by loved ones, women with secondary infertility are outlaws in the world of primary infertility. People with primary infertility feel that those with secondary infertility have nothing to complain about. Women with primary infertility would cut off their own arm to get one healthy baby, and thus they have very little sympathy for someone who has a child and is upset because she can't have a second or, God forbid, a third.

Even though they already have a child, women with secondary report just as much depression and anxiety as do those with primary infertility. This feels counterintuitive. You would think that a woman who already has a child wouldn't feel so depressed. You would think she'd be even less likely to be depressed if she went through infertility the first time around—that she'd be so happy just to have a child that she would be much better off psychologically than a woman with primary infertility. But that's not the case, because women with secondary infertility have a range of other secondary-only issues to deal with in addition to all of the usual problems of infertility.

Most people want their children to be two to four years apart so they can grow up and play together and be close. That expectation puts a lot of pressure on you—as every month goes by, you envision your children being further and further apart in age. You fear that they won't feel emotionally connected because of their age difference.

There's also a problem of logistics: If you're going through infertility treatment and you have a small child at home, what do you do with the child when you have to go in every day or every other day for vaginal ultrasounds and blood tests? I've had patients who had to bring their kids with them, and it's terribly stressful. People give you the evil eye in the waiting room, and often the child freaks out watching Mommy being probed or poked. But sometimes it's unavoidable. A lot of infertility treatment is last minute. The clinic calls and says your IUI is tomorrow, and you have to be there whether you can get a baby-sitter or not.

Secondary-infertility patients also find it hard to practice selective avoidance. People with primary infertility can live solely in the adult sphere—they can avoid anything that has to do with children. But if you already have a child, you can't escape. You have to walk into the preschool every day. You have to go into the toy store to buy presents for kids' birthday parties. You have to hang out with other women who are pregnant with second or third children. You can't distance yourself from it.

You may also worry about giving your kids a distorted view of reproduction. It's hard to hide from the child the fact that you're going through a high-tech cycle. If you're giving yourself shots twice a day, the chances that your child is going to walk in on you at some point during that time are pretty good. "What are you doing?" the child will ask. "Why are you having a shot? Are you sick, Mommy?" Those can be difficult questions to answer.

Financial concerns and issues of equity may arise as well. If insurance doesn't cover your infertility treatments, you may feel guilty spending money producing a second child when that money could be spent on your first. If you opt to have an IVF, will you be paying for it with your first child's college fund? Is that fair? In fact, a lot of women with secondary infertility don't even bother to seek treatment. They don't want to spend the money, and they don't want to risk having a multiple pregnancy.

You might even have a little person asking for a sister or brother, as my patient Carol, thirty-five did. Here's what went on in her life during secondary infertility:

Carol's Story

I had no difficulties whatsoever getting pregnant the first time. I got pregnant within a couple of months. It was a great pregnancy, a healthy child, everything. And I never expected to have trouble getting pregnant again. But when I started trying for a second child, I had problems. I conceived three times, but miscarried. Each time it took me longer to conceive, and after the third miscarriage I couldn't conceive at all.

It was extremely stressful, and I was very depressed. My son was in preschool, and everyone was having baby brothers and sisters, and they would make a big deal about it at school. He would come home asking me, "When are we going to get a baby?" He'd wake up crying about it in the middle of the night. I found that really hard.

I think the hardest part was that I was a stay-at-home mother and everyone around me was getting pregnant and having babies. Everywhere I looked, I saw families with kids. I used to sit in restaurants and look around and count how many kids everyone had. It used to drive my husband crazy when I'd say, "We're the only ones who have just one.

Everybody else has more than one." Then I would ask the other parents what their children's age differences were, and I would cry because I knew that my kids wouldn't be close in age. Every year at my son's birthday party, I would cry because he was another year older and we still didn't have another baby.

I really didn't want to have an only child. I had a real thing about that. I felt bad for my son that he was all alone—he would sit in the backseat of the car and say, "No one is back here with me, I'm all by myself." I was jealous of my friends who had two kids close in age, because they would play, and my son never had anyone to play with—I was his constant playmate. I also had an irrational fear that something would happen to him, and then we wouldn't have any *kids.*

When Carol joined my mind/body infertility group she was feeling "really freaky," she says. She felt very depressed, she had lost interest in the activities she enjoyed, and her marriage was suffering. Just making the commitment to come to class helped her feel better.

It was such a big thing for me to say, "Every Monday I'm going to do something for myself." I never did that before. Everything revolved around my husband's career and his job and his schedule. And for the first time he had to come home so I could do something for me. When he got home from work, I would bolt out of the house. I had a one-hour drive to get to class, and I loved that ride. It was just so great for me to get out and do something for myself.

Finally, with the help of drug treatments combined with IUI, Carol conceived. But the stress didn't end there—eighteen weeks into her pregnancy, tests showed that the fetus had a defect, although the extent of the problem could not be pinned down. A genetic counselor broached the possibility of ending the pregnancy.

We actually came very close to terminating the pregnancy because no one could really tell us what to expect, and it seemed like there were so many things going on with the ultrasound—there might be something wrong with the spinal cord or the kidneys, the heart or the brain—but they just weren't sure. At the last minute we decided not to terminate, and I'm so glad we made that decision. My son was born without toes on one foot, but otherwise he's perfectly fine—he's ahead of all the other kids in his play group, and he's reached all his developmental milestones.

I'd love to have another child, but I won't undergo any more infertility treatments. I'm worried that they might have been a factor in his birth defect. Maybe it will just happen naturally.

How do women like Carol survive secondary infertility? By packing their survival-skills toolbox with a special set of tools, including the following, designed specifically for secondary infertility.

Find Secondary Support

One of the best things you can do for yourself is to seek out the support of other women with secondary infertility. *But don't expect to get support from people with primary infertility.* If you join a group with women who've never had a baby, you may not feel welcome. It's hard for them to understand what you're going through. Instead, seek out people with secondary infertility. Ask your doctor if he or she has other secondary patients you can talk to, or see if your local Resolve chapter has a secondary-infertility support group. If neither of these approaches pans out, you may have to be creative: Make friends with the woman in your doctor's waiting room who has a child in tow. Look around the playground—if you see with someone with a five-year-old with no siblings, there's a chance that mom has secondary infertility and feels as isolated and desperate as you. Of course, you can't just walk up to her and say, "Hey, do you have secondary?" But you can start a conversation and try to lead it to infertility. For example, if you see a pregnant woman, say something like "I wish that were me," and see what she says. You'd be amazed—women with secondary infertility are all over the place.

Watch for Depressive Symptoms

Don't think that just because you already have a child, you can't become depressed over your infertility. I've seen patients who become clinically depressed because they can't get pregnant with their third child. One patient of mine is distraught over not being able to give birth for a fifth time! From what I've seen in my clinical practice, it's not the number of children you have that matters, *it's not having the number you want*. Anytime anyone is denied what they want, it can become a crisis that triggers depression.

Secondary patients report every bit as much depression and anxiety as do women with primary infertility. They don't want just one child—they want a family. They may envision having a large family and can't fathom having only one child. Or perhaps they love their singleton so much they feel they'll burst if they don't have an additional child to love. Whatever the motivation, please keep in mind that whether you're pining for your first child or your fifth, you may be depressed, and if you're depressed, you need professional help.

Be Mindful of Number One

A lot of the patients tell me that they feel as if they're missing their first child's childhood because they're so focused on getting pregnant the second time. I think that is really tragic. I see patients with primary infertility who get pregnant and literally start worrying about conceiving a second time as soon as they find out the first pregnancy is viable. I've had patients who stopped nursing their babies at six weeks of age so they can start another IVF cycle.

I'm not saying you shouldn't try for a second child, but I am urging you not to do so at the peril of your first. Be mindful of your child's milestones—her first smile, his first step, even her first temper tantrum. These are all such precious memories, and if you are so obsessed with getting pregnant again, you may forget to enjoy the baby or child that is already in your life.

Acknowledge the Validity of Your Feelings

Other people may say things that cause you to doubt yourself. "You have one baby, isn't that enough?" or "You're lucky even to have one—isn't it a little selfish to be upset that you don't have two?" When you hear comments like this, try to push them aside. Recognize that your feelings of secondary infertility are normal, and you shouldn't beat up on yourself for having them. Of course you're grateful to have one child, but you want one more. Or two or three or four. Even though you already have a child, your feelings of loss and anger and frustration and depression and anxiety are just as legitimate as they would be if you had primary infertility. They may be different, but they're just as valid.

Assert Your Right to Relax

It can be extremely hard to carve out time every day for relaxation when you already have a child. If you're a stay-at-home mother, you're occupied with child-care duties, and if you work outside the home, you're juggling work and commuting and child-care. Either way you may feel uncomfortable asking for time to spend on yourself.

Don't feel guilty. You deserve relaxation time. It will help you cope with the emotional upheaval of infertility, and it may even help you conceive.

Of course, finding time can be a challenge. If possible, ask your husband or your mother or someone else to watch your child while you do your relaxation exercise each day. If your husband balks, tell him that the whole family will have a much more pleasant evening if you have time to relax. If that's not possible, try these:

- Take a mindful walk while your child sleeps in the stroller.
- Relax during your child's nap—the laundry can wait.
- Take time for yourself after your child goes to bed. Again, you may have to put off household chores or accept the idea that your house may not be as clean and tidy as you'd like it to be. Don't sweat it. Relaxation and self-care are more important than a clean house.

- Swap time with another parent—you watch her child for an hour, then she watches yours.

Challenge Your Thoughts

If you're thinking things such as "I can't get pregnant again because I'm not a good enough mother" or "I must not like being a mother" or "My child will be harmed forever if I can't produce a sibling," you need to spend some time restructuring your thoughts. Put them to the test—are they valid? Are they logical? How can you restructure them in a more constructive way?

Write It Down

One of the things that bothers secondary-infertility women most is that they don't want their child to see them cry. But this is hard—you get your period or your doctor calls with bad test results, and you have to hold it all in because your child is with you.

Journaling can help in such situations. If you can't talk to somebody, sit down and write about it in your journal. It's a great way to let go of some of your feelings without losing it in front of your child.

Do Millions of Minis

Another way to stay in control in front of Junior is to do a mini. A mini can hold you together just long enough to get him in front of the TV, turn on a video, and run to the bathroom, where you can let it out in private.

Educate Yourself on Single Childhood

You're dead set against having an only. But are your assumptions about single childhood true? To be sure, educate yourself on single childhood. Read books, talk with parents of only children, talk with friends without siblings. Is it as bad as you think? Research on single children says that, despite the stereotype, only children are not selfish, egocentric, neurotic people. In fact, studies say that only children do very well. They're well adjusted, they're high achievers, and they're comfortable around adults. In France, according to a patient of mine who suffered from secondary infertility and wrestled with the idea of being content with only one child, only children are referred to with the lovely name of *enfants uniques*. If you're pressuring yourself to have a second child just so your first child won't be a spoiled brat, you may want to reconsider your reasoning.

Think About What to Say to Number One

Taking into account your child's age, personality, and curiosity level, work with your husband to formulate a plan on what you will and will not tell your child regarding your infertility treatment. Some decide not to tell their children anything, and others are completely open. It depends on your situation. My patient Olivia decided to be completely open with her four-year-old daughter, Megan.

> *Darren and I leveled with our daughter. We told her,* "Mommy needs some medicine to help her have a baby." *We made it a family affair. When Darren had to give me shots, my daughter would wipe my skin with an alcohol swab and hold my hand while Darren gave me the shot. We did this twice a day. After about a week of this,* Megan said to me, "Mommy, *when I have a baby will you hold my hand during my shots?" That was bittersweet—she was being so supportive of me, but it's given her a very different view of how babies are made. I hope she doesn't have to go through what I've gone through.*

Not every family is that open with their kids about infertility treatments. Some parents just tell them that they'd like to give them a brother or sister and they are going to a special doctor to help get a baby. It all depends on you, your child, and the degree of openness with which everyone is comfortable.

 Infertility in Unmarried Women

As a single woman or lesbian who can't get pregnant, you are not only struggling with the agony of being unable to conceive, but you're more than likely dealing with family, friends, co-workers, and perhaps even a doctor who may not fully support your decision to become a parent in the first place, let alone to pursue infertility treatment. And what's worse, if you're single you're probably trying to cope with all this tension alone.

Quite a few single women, both straight and gay, have participated in my mind/body infertility program. I believe that the program has helped them because it gives them the tools they need to develop the inner strength, the dogged perseverance, and the emotional flexibility they need to cope successfully with the extraordinary challenges of being unmarried and infertile.

The best way for me to begin this chapter is to you tell you about Angie, a thirty-eight-year-old office worker, whose story sums up so many of the problems faced by unmarried infertile women.

Angie's Story

Angie always assumed that her life would take a typical path: She'd go to college, get a job, get married, and have children. But life didn't turn out the way she expected. After finishing school and establishing her career, Angie started looking around for a husband. She met some nice guys, but never Mr. Right. As the years passed, Angie started to worry—what if she didn't get married? What would happen to her dream of having children? She couldn't even imagine a life without a family. At age thirty-five, Angie decided that if she were not married by the time she turned forty, she would adopt a child and be a single mother.

Unfortunately, that plan left her feeling uneasy.

"I didn't have any peace with that decision," says Angie. "I realized that I just didn't want to pass up the experience of being pregnant and

giving birth. So I decided when I was thirty-seven that I was going to do this on my own." Angie chose an anonymous sperm donor and began having IUIs at a local clinic.

Nine IUI cycles and three donors later, Angie still wasn't pregnant. "It became clear to me that IUIs were just not going to work and that it was time to consider IVF," Angie says. She was devastated. "I was so angry, so incredulous that I was going to have to go to these lengths to have a baby. It was bad enough that I didn't have a husband. But to have to resort to IVF—I just couldn't believe it."

Angie's IVF failed. Now she is deciding whether to have a second IVF or adopt. It's not an easy decision, especially since she feels so reluctant to give up on being a biological parent. "All along I've had to give things up—starting with a husband. Now I'm struggling with the idea of giving up being genetically connected to a child. I never imagined that I would do this in any other way but the traditional way—have a husband, have a family. I got exasperated with my shrink last week when I was talking with him about the possibility of using a donor embryo. He said, 'You know, you may have to go to Plan B.' Well, I almost lunged across the office and grabbed him by the neck. I said, 'Plan B? That was last year. I went to Plan B when I decided to be a single parent. I'm on Plan X now. There's just Y and Z left!"

Angie's situation is further complicated by the fact that her health insurance doesn't cover IVF. She used a home-equity loan to pay for the first one, which cost eleven thousand dollars, and money is on her mind as she decides what to do next. "Do I spend all my money on medical treatments and leave myself no money to adopt? I don't want to do that," Angie says.

Making these decisions alone has not been easy. "I have no one to bounce all this off of. It truly is my decision. Also, because I'm single, I have to do all the arranging—I have to hire a sedan service to drive me to and from the IVF egg retrieval and embryo transfer, and I had to ask friends and neighbors to help out when I spent three days in bed after the transfer—they brought me food, fed the cats, everything. There are so many logistical things to deal with, and that adds more stress.

"A lot of infertile women talk about feeling isolated. Well, I feel isolated times two, because not only am I doing high-tech medical treatments to try to get pregnant, but I'm doing them alone. I haven't told a *lot of people that I'm even trying, because I didn't want to get into the judgments. I don't want this to be a moral issue." Although her mother, a Catholic, has been supportive, Angie hasn't discussed her parenthood plans with her siblings, who are rather conservative. She found some understanding friends in a support group for infertile women, but she is the only single group member, which sometimes leaves her feeling left out.*

Angie knows only one other single infertile woman—and she lives in Asia, of all places. "The nurse in my clinic hooked me up with her. She travels to America for infertility treatments. We had dinner one night about six months ago when she was in town, and we keep in touch by e-mail. She's been a great source of support for me, and I hope I have been for her, too. I just wish she didn't live so far away. It would be nice to have someone a little closer to home."

People never really think of single women as going through infertility. You hear about single women getting pregnant by accident or by choice, but rarely do you hear about a single woman who is trying desperately to get pregnant. And yet there are so many women who tell themselves that if they don't get married by a certain age, they're going to have a baby on their own. They wait and wait and wait to find a husband, but eventually they give up and choose single parenthood. Since these women are usually in their mid- to late thirties or early forties when they begin trying to conceive, they are at greater risk of infertility than they would have been in their twenties or early thirties.

Unfortunately for single women, the infertile world is very much set up for the couple—peek into the waiting room of just about any infertility clinic, and you'll see a Noah's Ark–like lineup of couples, two by two. In fact, a lot of infertility doctors won't even treat single women or lesbians—they think that only married heterosexual couples should have babies. So you might have some difficulty even finding a doctor who will treat you. A lot of physicians are very traditional, and unless your physician seems very supportive of you as a single woman, you may not feel that he or she is fully invested in your care. That's a problem, because if you don't trust your physician completely and you fail to conceive, you may spend the rest of your life questioning his or her treatment recommendations. That's not how you want to live.

Friends and family may offer little help. They may criticize your decisions, say you're crazy even to think about raising a child alone, or tell you that it's unfair to bring a child into the world without a father. When you suffer a setback in your quest to become pregnant—a negative pregnancy test or news of elevated FSH results, for example—you may have no one to turn to. It can be a very lonely time, and it can leave you feeling like a failure. Not only are all your friends married, but they're having babies, too. You can't succeed at either.

Lesbian women have an additional set of challenges, although if they're in a committed relationship with another woman, they're more likely to have the loving support of a partner. Their families may not accept their nontraditional lifestyle or their decision to bring a child into a gay family. Even though research shows that children of gay parents fare as well as children of heterosexual couples, unenlightened family members often feel that a gay household would be an unsavory place for a child to grow up. Family members may feel that by supporting a gay woman's quest for a child, they would be endorsing a lifestyle of which they don't approve.

Whether you're gay or straight, you may face donor troubles. When you first decided to get pregnant, you formulated a plan on who would father the child—either someone you know or an anonymous donor. If a friend agreed to provide you with sperm, his commitment to the process may be tested by infertility. When you first started trying to conceive, he probably assumed you'd get pregnant quickly. But will he stay committed to the process if it takes months or years? Is he willing to have a medical workup that will measure his reproductive health? Is he willing to drop everything at a moment's notice when you ovulate in order to provide sperm for a treatment? If he agreed to provide coparenting or financial support to the child, how would he feel if you had twins or triplets?

For that matter, how would *you* feel, as a single mother, about having twins or triplets? Raising multiples is a challenge even for married couples with loads of family support. Do you have the social and financial resources to raise more than one child?

Often men who agree to father a child for a single woman are not interested in participating in round after round of infertility treatment.

They simply don't want to be that involved. Many single women who start off trying to conceive with a friend's sperm eventually resort to donor sperm.

Financial issues also loom large for single women. Many have health insurance that does not cover infertility treatments. Indeed, some even have trouble getting their insurance company to acknowledge their infertility. This happened to Angie. "At first I couldn't even qualify as being infertile, because I was single, and the criterion for infertility is failing to get pregnant after six months to a year of unprotected intercourse. I wasn't having sex with anyone at that time, and besides, unprotected sex wouldn't be safe for me as a single woman! But I begged the insurance company and said I'd had six unsuccessful IUIs—isn't that worth something? Eventually they gave in, and I got my infertility diagnosis, and they covered the last three IUIs, which were about six hundred a month. But they wouldn't cover IVFs; those are just not allowed in the policy I have." Faced with paying for IVF herself, Angie requested—and received—a 20 percent discount from her infertility clinic. Some clinics offer reduced rates to patients whose income falls below certain levels, and single women are more likely to qualify for these rates than are couples with two incomes.

COPING SKILLS

Because of the many difficulties presented by single infertility, it is absolutely crucial for single women to develop a wide range of infertility coping skills and strategies. The following are among the most helpful:

Develop a Network of Supporters

The most important coping tool for the infertile single woman's toolbox is social support. I can tell you unequivocally that a strong network of social support makes single infertility far less distressing than it is if you are isolated. Start by connecting with other single infertile women. Some cities have established support groups—check with Resolve or your infertility clinic to see if there's one in your area. If there isn't, create your own. This may not be easy, but it is worth the effort. Post notices on bulletin boards, search among Internet chat groups, ask your health-care providers for leads. Do whatever you have to do to find at least one other single infertile woman with whom you can share your experience on a day-by-day basis.

Look for other friends you can rely on, too. If you have friends who have supported you so far, sit down with them and ask them to make a firm commitment to help you through your infertility. Tell them what you need—someone to accompany you for treatments, someone to hold your hand after an unsuccessful cycle, whatever. If you're lucky, you'll find a friend or two who can really step up to the plate and be there for you. And remember, your need for a strong social network won't end when you become pregnant. As a single parent, you'll have to rely on friends and family far more than you would if you were married. Build those relationships now, and nurture them so that they'll last throughout your life.

If you have friends or family that are not supportive of your choices, cut them out of the conception process. Don't share details about which procedure you're having next week or what step you'll take if that fails. You don't need the pain of their disapproval; it will only add to your distress. Save your energy for fostering positive relationships, not coping with negative ones. But don't give up on your family completely—they may change their tune after you have a baby. Most of my single patients have told me that once their baby arrives, their families' hard opinions soften and they welcome the child into the family. They may not approve of the fact that there's no father, but their excitement at having a new member of the family outweighs their disapproval.

See a Therapist

As I discussed in Chapter 1, infertile women are at increased risk for depression. Single women are at even greater risk for two reasons: First, they often face greater challenges and have less support than married women do. Second, they have no spouse to detect and point out that they are depressed. A married woman might seek help for depression because her husband begs her to do so. A single woman's depression may go unnoticed, or a friend or family member may feel reluctant to urge counseling. A therapist will help you determine whether you are depressed and, if you are, how best to treat your depression. Remember, you can't necessarily depend on your own instinct when it comes to deciding whether you are depressed, because depression and denial often walk hand in hand.

A therapist can help in other ways, too: with decision making, building coping skills, negotiating family problems, and dealing with disappointment and loss.

Keep a Journal

One of the most difficult parts of being a single infertile woman is making decisions. When should you pursue high-tech treatment? How do you choose a sperm donor? Which doctor should you pick? How do you tell your family that the father of your child will be a gay friend at work? These are all incredibly difficult choices, and if you're making them alone, they're even tougher.

Journaling can make them a little easier. Many women find that when they write about their choices, solutions sometimes present themselves more readily than when they're just thinking. In a journal you can list the pros and cons of a decision, explore fears and worries, walk through what will happen if you do this or that, and vent any anger or frustration.

Do Relaxation Exercises Daily

Single infertile women generally have more distress than do their married counterparts. Not only are they facing infertility, but they have a host of additional worries that result from their singleness. That's why relaxation is so important. Make sure to do some kind of relaxation session every day, or twice a day if you're very anxious.

Restructure Your Thoughts

Any single woman pursuing infertility treatments will have doubts and second thoughts about her choices: "Oh, my goodness, I'm not going to be able to do this by myself" or "It's unfair of me to bring a child into the world without a father" or "My family is right, I shouldn't be doing this crazy thing!"

Cognitive restructuring will help you take the punch out of negative thoughts. It will also help you make sure that what you are thinking is rational and logical.

Educate and Communicate with Your Donor

If a friend is serving as your donor, make sure he knows what infertility requires of him. Talk with him about what he can expect from the process, including the scheduling demands of procedures such as IVF, and make sure he is willing to stay committed. It may help for the two of you to meet with an infertility counselor, who will see to it that your donor understands everything that will be needed of him. Although he may run for the hills when he finds out the details of infertility treatment, it's better that he backs out early in the process rather than, say, in the middle of a cycle.

Discuss details such as scheduling, financial responsibilities (who will pay for his medical workups?), multiple births, and what presence, if any, he will have in the life of your child (or children).

Consult a Financial Adviser

What path you take in treating your infertility may depend on your financial situation. How much can you afford to spend? Does a second mortgage or a loan make sense? How much of your savings do you need to hold on to so that you're not completely broke *after* your child arrives? Consider sitting down with some kind of expert who can help you decide what you can and can't afford, and how you can use existing investments to pay for infertility treatment or adoption. A financial adviser may also be able to help you figure out how to finance a leave of absence from work, if that's what you'd like to do.

Be Mindful of What You Do Have

You don't have a husband. You don't have a child. You may not have a supportive family. As a single woman, it's extremely easy to ruminate on all the things you *don't* have. But being mindful of what you *do* have can be very helpful. Think in terms of both infertility (you do have the freedom to make decisions without another person's disagreeing) and everyday life (you love your job, you have wonderful friends, you live in a cozy cabin in the woods that allows you endless hours of communing with nature). Being mindful of the good things in your life that can help take away a little of the pain of infertility.

Nurture Yourself

As a single woman, it's your job to make all the infertility decisions. Plus, you have to do everything else, including earning money, paying bills, taking care of the house or apartment, maintaining your car, and so on. It's easy to get so wrapped up in the everyday responsibilities of life that you forget to nurture yourself. Don't fall into that trap. Unwind each day by doing something nice for yourself for at least half an hour. Choose something that fulfills you, brings you joy, and takes your mind off the responsibilities of infertility. Anything from a nap in a hammock to a night at the opera qualifies—just make sure to take care of yourself in a loving, joyful way.

CHAPTER EIGHT

Why Won't God Give Me a Baby?

I've had a real crisis of faith during the past two years. I pray and pray, and it just doesn't seem to work. I don't know why God isn't answering my prayers.

I grew up Catholic, and I'm comfortable with it, although I don't agree with all the rules. Birth control is a perfect example—of course I took birth control; I took the pill for ten years when I was single. I wasn't going to wait to have sex, but I also didn't want to be a parent with someone I wasn't married to. Somehow I made peace with that, and I could still think of myself as a Catholic but do what I needed to do. But now I wonder if I'm being punished for that.

My patient Amy has unexplained infertility, and her guilt was compounded when she began infertility treatments, some of which the Catholic Church forbids. She had several IUIs and one IVF, and they all failed.

I still go to church, but less and less often. I'm just so torn, and I'm not sure what to do. This is not something I would talk to a priest about,

> *because I'm sure he would say, "Of course God isn't answering your prayers—you shouldn't be doing this! Go home and say the rosary, and don't ever go to the infertility clinic again."*

Infertility is the first experience in many women's lives when God and religion don't seem to have the answers. They beg God to let them conceive and carry a pregnancy to term, but months and years go by without a baby. They feel that God is punishing or abandoning them, or that infertility is proof that there is no God. They wonder if God is preventing conception or causing miscarriages because God feels they won't be good mothers. Doubting God for the first time in their lives, many begin to question the faith they grew up with. And of course that makes them feel guilty.

All this questioning leads women in several different directions. Some lose their faith completely. Others survive the crisis but emerge with a changed view of their religion or with a completely different kind of faith. Still others find their existing faith strengthened by the trial of infertility and move to a deeper, more spiritual place.

How can infertile women resolve the religious doubts that arise when their prayers go unanswered? How can you negotiate your way through a crisis of faith and emerge with a stronger, more mature relationship with God? Can marginally spiritual women who are drawn to God learn to derive strength from their faith during their struggle with infertility? And how can you learn to accept infertility as God's will without being angry at God for choosing a fate that is so far away from what you would choose for yourself? How can you find spiritual peace when your world is in a state of upheaval?

These are questions infertile women grapple with, and they are questions my patients ask me all the time. I'm no spirituality expert, though. That's why, both in my mind/body infertility program and in this chapter, I turn to someone who *is* an expert in this area. The Reverend Dr. Barbara Nielsen, an Episcopal minister and psychologist, has been doing spiritual and psychological counseling in the Boston area since 1978. Barbara worked with us at the Mind/Body Medical Institute for two years while she was getting her Ph.D., and during that time my patients and I were truly blessed by her knowledge, insight, and sensitivity in the area of infertility and spirituality. Barbara helped many of my patients navigate their way through the rockiest of spiritual crises. She inspires infertile women to explore their feelings and find answers that are right for them. She also advises them on how to use their infertility as an opportunity to move gracefully to a more mature, more enlightened level of spirituality.

In this chapter I'm going to discuss how infertility challenges a woman's spirituality and how she can cope with and move beyond these challenges. Some of the insights and suggestions will be mine, based on what I've learned from years of counseling infertility patients. But the lion's share come from Barbara. I thank her for providing them.

First, a note about my word choices in this chapter. When I write about God, I am referring to the entity that many mainstream religions consider to be the Creator of the world and Supreme Being. And when I say "church," I am speaking in a generic sense about a place in which a faith community comes together to worship. Those words mean different things to different people, however, and I encourage you to replace them with words you're more comfortable with, if you wish. My comments here refer to fairly conventional, traditional religious beliefs, because those are the beliefs most of my patients hold. If my discussion lacks relevance to your own faith, I do hope you will find some benefit in this chapter nonetheless.

Spiritual Stages

When it comes to spirituality, many people exist in a "conventional" stage of thinking. As we grow from childhood to adulthood, we develop a particular way of looking at ourselves and our world and our religion. This unique worldview reflects what we learn from our parents, our places of worship, our schools, and our communities. A person who is in the "conventional" stage of spirituality essentially accepts what she's been taught, give or take a few rules here and there. She attends religious services sometimes but not always. She has faith because she's been taught that having faith is the right thing to do, and it feels comfortable to her. But her faith has not been tested, and she has not actively made it her own.

When a major life upheaval strikes, conventional thinking often falls apart. Untested faith lacks the strength to sustain you through trauma or catastrophe or crisis. The tenets that sustained you during the less tumultuous parts of your life can collapse. You may begin to question God, your beliefs, the rules of your religion. Everything begins to break apart.

Thus begins the "critical" stage of spiritual thinking, the stage in which many infertile women find themselves for the very first time. As a woman goes through infertility, she is challenged at the deepest level of her being. As she questions and examines and challenges her own conventional thinking, she thinks critically and makes decisions about her beliefs. Eventually she discards what is of no use to her and embraces what she truly believes. Make no mistake, it is a difficult process to let go of old ideas and ways of being that have served you well for so many years, as it was for my patient Bette:

> *I always felt like I had a lot of faith. Then, with infertility, I was beginning to question that faith. I'd always felt that if you had faith, you trusted in God to let your life unfold the way it is supposed to, and bad things that happened were your lessons to be learned. But when it came down to realizing that this might be the hand that I would be dealt—that I wouldn't have children—I started getting pretty angry and feeling like I'd gotten a raw deal. That was really hard for me. Not only was I not having my children, but I was also being faced with losing the beliefs that I'd held all my life, as well. I felt like I'd totally lost control of my life—everything was not what I had thought it was.*

It can be painful to take on new ideas, particularly when they go against decades of familiarity or family tradition. People who are raised in rigid religions with lots of rules and strict codes of behavior often have the most difficulty, because they feel that the rug has been pulled out from under them. But no matter how arduous it is, this is a normal, natural process into which most spiritually minded adults are eventually pushed.

It's easy to blame infertility for the forced march from conventional to critical thinking about faith, but placing blame might be a mistake. If infertility doesn't push you to think critically about faith, sooner or later something else will—a cancer diagnosis, the death of a loved one, a family tragedy, or some other crisis. Instead of pointing fingers at infertility for wrecking your relationship with your Creator, try to accept it as something that would have happened eventually anyhow and consider it an opportunity to grow and develop your spirituality and to move to a deeper stage of being. Many women turn infertility's crisis of faith into a path to enlightenment.

But believe me, it's not simple.

Any journey is easier when you have the loving guidance and support of an experienced traveler. This is especially true with a spiritual journey. That's why most religions have rabbis, priests, ministers, or some kind of spiritual leader to assist seekers as they work to move themselves closer to God. Unfortunately, as infertile women tackle the spiritual questions that critical thinking uncovers, as they set out to negotiate their own spiritual crisis, they often have nobody to guide them. It's sad but true: When it comes to infertility, organized religious communities often just don't get it. In fact, they frequently do more harm than good to an infertile couple's aching soul.

In a perfect world a woman struggling with infertility would be able to turn to her spiritual community for love, support, empathy, and understanding. But, ironically, those things are often in short supply in religions that put the family at the center of their universe. They understand family, but they don't necessarily understand the anguish that one of its members may feel when God doesn't give her children of her own. One of the hardest things for infertile women who are members of churches or synagogues is the emphasis that most churches place on parenting children. During wedding ceremonies, Bible readings urge brides and grooms to go forth and multiply. Some religions, including Catholicism, Orthodox Judaism, and the Church of Jesus Christ of Latter-Day Saints, encourage their members to have large families. And worship services of many faiths are often packed with children, particularly on holidays. This can cause an infertile couple enormous pain, as it did my patient Amy.

I found out on Good Friday that my IVF had failed. Then, on Easter Sunday, I went to church. It was the worst thing I could have done. Everywhere I looked, there were families and kids and babies in Easter bonnets. I know I shouldn't idealize other people's lives, but I couldn't help it that day. All the prayers made me cry. I just went home and took to my couch for the rest of the day.

There seems to be no place for the married couple without children in some religious communities. Ironically, however, some of these very same faiths forbid the medical treatments that can help turn an infertile couple into parents. The Catholic Church forbids IVF and other infertility procedures. Orthodox Jews and Catholics are not supposed to masturbate—an annoying but necessary step in infertility treatment. (Note: Special sperm-collecting condoms can be used to get around this rule.) Some fundamentalist Christian churches frown on egg or sperm donation because it introduces a third person into the sacred union of two married people. So what happens here is that you're told to have children, but when you can't, your religion can do nothing for you. You're on your own.

Infertile couples often feel that it is hopeless to ask a clergyperson for advice or support, particularly if their religion forbids infertility treatment. As one of my patients said, "What the hell does a priest know about infertility? An unmarried man would never understand my need to have a child." As a result, infertile couples can have a hard time finding support, and in time they become angry and isolated from their faith community.

Both Barbara and I believe that it's worth it for infertile couples to take the bull by the horns in situations like this. If you're struggling with a crisis of faith and you need help, don't sit back and assume there's none out there for you. There may be. But you're going to have to search for it, just as you may have had to search for the right doctor, supportive friends, understanding family members, and the most effective communication skills for you and your husband. No, it's not fair that you have to do all this searching. But it's usually worthwhile.

Asking for Spiritual Support

A while back Barbara set out to find clergy members in the Boston area with experience counseling infertile women. She didn't find many, but there definitely were some clergy with a personal or professional interest in infertility. Among her discoveries was a rabbi who had experienced infertility and who had a tremendous amount of empathy to offer the infertile couples he counseled. She also came across some Catholic spiritual directors who were helpful.

We recommend that even if you think your faith community's leaders won't be able to help you, give them a chance. It's worth a try. Yes, you may strike out, but you may not—maybe you'll discover a clergyperson who does understand your situation, either because of personal experience or because he or she happens to draw from a very rich, deep well of emotional understanding. Be ready to look beyond the clergy, to other members of the church staff. Perhaps a nun, a ministerial assistant, a rabbinical student, a visiting scholar, or a very involved layperson within your faith community may be able to offer guidance.

Ask around—have any of the women in your infertility support group come across a minister or rabbi who's been particularly helpful? Perhaps that individual would be willing to counsel you, even if you're of a different faith. And if you can find one, a spiritual counselor might also help you find answers—sometimes an individual will leave a religious order and work as a spiritual counselor. These counselors are trained in matters of the soul but don't necessarily have ties to a particular institution or set of rules. Unfortunately, spiritual counselors can be difficult to find—my only advice is to seek and perhaps you will find one. Hospitals often have lists of pastoral counselors in the community they serve. And check with your therapist, if you're seeing one. Perhaps she has experience with spiritual counseling that you don't know about, or she may be able to recommend someone else to whom you can talk.

When you approach your clergyperson, lay your needs out on the table—you're struggling with infertility, and the struggle is testing your faith. Ask, "Can you help me? If not, do you know someone who can?"

When you meet with your clergyperson to discuss your situation, bring along educational materials about infertility. Who knows? Maybe you're the first infertile couple he (or she, but I will use "he" here to avoid awkward sentences) has met, and he doesn't know the issues involved. If a clergyperson is serious about doing a good job, he will welcome the opportunity to learn about a problem that is affecting the spiritual community. You may have to bring him up on the infertility learning curve, but don't let that stop you. Just as your spouse, your family, your friends, and your co-workers don't know how exactly to nurture you during this time unless you tell them what you need, neither will your priest, rabbi, or minister. Give it a try. And if he has no skill or interest in helping you, move on. Maybe someone else can.

Finding a New Faith Community

Sometimes as infertile women search their souls for answers during their critical thinking about spirituality, they discover that they are repelled by their old, comfortable religion and feel attracted to a new religious denomination, church, or faith instead. This can be a little unnerving, particularly if you've followed one tradition for many years. But Barbara believes very strongly that you should be open to the idea of moving to a new spiritual community if you outgrow your current one, and I agree wholeheartedly. If you're not getting what you need, look around. Attend other churches. Sample other denominations. It can be a very liberating experience. Start by attending a few services and social activities at the new place. As you do, ask yourself these things:

- What's the community like? Is it friendly, outgoing, and open? Is it accepting of new members?
- What kind of religious education is available for adults? Are there formal classes that give both newcomers and current members a way to learn about, explore, and discover their faith?
- How much work does it do within its own community? Are

there prayer groups? Support groups? Seminars? Are the needs of the sick, the dying, and those in crisis attended to in an effective, organized way?

- What does it do in terms of outreach to the community—the neighborhood, the local area, and the world? Do members raise money for international tragedies? Collect food and clothes for inner-city churches? Does it encourage cross-culturalism?

After those questions are satisfied, talk to the head clergyperson. Ask, "What's your vision for this spiritual community? What are the important things it stands for? How are those priorities reflected in weekly services and liturgy? How does your church respond to the needs of individuals?" Explain your own situation, and ask, "What can you do to help me?"

Don't be afraid to shop for a place of worship the way you would anything else that's important to you—a new house, a new car, a caretaker for an elderly relative. Participating in a particular faith community just because it happens to be close to your house or because you've been going there for years doesn't make sense, especially if your spiritual needs are not being met. As we change and grow, our spiritual needs change and grow, too.

Making Peace with Punishment

Do you feel that God is punishing you with infertility? Many of my patients do. They believe God is angry at them for having had an abortion, or for living with a partner to whom they are not married, or for having multiple sexual partners. The idea of an angry, vindictive God is very present in some religions. Those traditions interpret God as telling you what rules to follow, and if you don't do just exactly what God says, you'll get it and you'll get it good. If that's what you believe, I can certainly see why you think infertility is a punishment, as my patient Carol does.

I definitely felt when I was going through infertility and my miscarriages that I was being punished for something. Something that I must have

> *done, although I don't know what. There must be some reason I couldn't have a baby, and I kept asking myself, what could the reason be? I must have done something that caused this. You rack your brains trying to figure it out.*

Many of my patients believe that infertility is a punishment. But I don't buy it. I argue against that idea on two fronts: First of all, do you think that infertile women are the only ones who have sinned? Didn't any of your fertile friends have abortions and premarital sex and affairs and do all kinds of other "bad" things? If you're being punished, why aren't they?

Second—and this is really the stronger argument—if you believe in a vindictive God, maybe you should be going back to the Bible or the Koran or the Talmud or whatever sacred text your faith looks to for answers and take a look at who your God really is. If you're seeing only God's vindictive side, I believe you're looking at God with one eye closed. I can't think of any faith that claims a God who does not love people and who will not forgive their mistakes. Ask yourself this: Doesn't my God promise love? Aren't we made in God's image? It's really hard for some people—women especially—to accept the idea of unconditional love. But that's what most faiths claim to have, a God who loves unconditionally.

That doesn't mean God ignores sin. But think about it: God doesn't expect you to be perfect, but rather to be in the process of perfecting your soul. The story in the Bible of Adam and Eve in the Garden of Eden explains to us the struggle all people have with putting our will above God's will for us. Mistakes are necessary for learning and for growing—God knows that. If you've done something that you and your faith consider sinful, and it is weighing on your conscience, seek forgiveness. Forgiveness is a fundamental part of just about every religion—I can't think of any religion that has no process for its members to ask for absolution, either via a clergyperson or directly from God. If you believe that God is punishing you for something, and you truly regret whatever that is, ask for forgiveness. Do penance, if that's what your faith requires or recommends. Then let it go. And if you don't regret your "sin," talk it through with God. Tell God why you did what you did. Ask for understanding. Ask for peace. Then listen for God's healing answers.

Cognitive restructuring works here, too. For example, maybe you think God is punishing you for having had premarital sex. Try to restructure that thought. Does it make sense? You could try to restructure it in this way: "It was a blessing that God sent that partner to me. I was smart to test him out before marriage, because marriage is a big step, and I didn't want to make a mistake. I did what I thought was right for me, and I stand by that choice. And I believe that God understands that. I am infertile not because God is punishing me but because God has a plan for me that does not include a baby at this moment." Restructuring your negative thoughts can help you eliminate the guilt and the self-flagellation and free you from the emotional destruction such thoughts can carry, as it did for my patient Kim.

> *I really thought that God was punishing me for premarital sex, and finally I composed myself and said, "Oh, please. I've had six partners in my life. I'm not a slut." There are plenty of people who have done things that are a lot worse. I had to stop thinking of it as punishment. Now these thoughts come back sometimes, but not as strongly as they did before. It helps me to think of success in the area of pregnancy as just being random, as opposed to there being some kind of merit-based system for motherhood, and I'm not meeting the criteria. That's irrational. I just read in the paper about a twenty-year-old who had her baby in a motel room and wrapped it in a newspaper and left it to die in a Dumpster. When I hear things like that, I think, "She managed to get pregnant. This just can't be merit based. It's got to be random."*

Mind/Body Prayer

If you feel that God is not answering your prayers, it might be worth taking a look at how you pray and perhaps trying a new way—a mind/body way.

Back in Chapter 2 I described prayerful meditation. Many of my patients have found prayerful meditation to be an incredibly useful way to pray and to prepare for prayer. There are two reasons for this, aside from the many benefits of eliciting the relaxation response: First, the meditative part of prayerful meditation allows you to slow down, calm down, and chill out in such a way that your mind is truly ready for prayer. Second, by slowing down and emptying your mind of other thoughts, you are much better able to listen to what God has to say after you present God with your requests.

As Barbara often says, so many of us, in our relationships with God, ask and ask and ask—but we don't stop to listen for an answer. Prayerful meditation gives us time to listen. If you're asking God to let you become pregnant, and it isn't happening, maybe God is whispering an explanation and you're not hearing it—either because you're so wound up in everyday life or because you don't *want* to hear it.

Barbara recommends praying meditatively. Ask and listen, ask and listen. Then end your prayer with the words "Thy will be done."

This is hard to do, especially if you suspect that God's will is different from yours. But by putting your fate in the hands of God, you are making a faith statement. You're trusting that God knows more than you do and that the outcome—even if it doesn't involve pregnancy and childbirth—will somehow be better, because it is God's will and God knows best. God knows the whole picture, while we have only our particular view. This requires enormous faith. For some women it is the ultimate test of faith and spiritual strength. And for others it is a relief to let go—to accept that their fate is out of their hands, as it was for my patient Gwen, who was not especially religious before her experience with infertility began.

> *Infertility has actually made me grasp on much more to praying and meditation and knowing that I'm not in control. There are some things I can do to achieve my goal of getting pregnant, but there are a lot of things that are not in my control, and I'm really powerless over a great deal of this process. I just have to turn it over and believe that the end results will be what they're meant to be. I can accept that and live with that.*

How do you develop this faith, this strength? How do you learn to accept the life path that has been chosen for you? It's not easy. That's where prayer comes in. Faith is a gift, in my opinion, and you can't create it. You ask for it, and God gives it to you. You can't invent it, even if you try, and asking God for assistance helps, as my patient Ellen concluded.

> *I was angry at God for making me infertile, but I came to terms with my anger in a couple of ways. I talked to a priest at my parish, and he was actually quite good. I told him that I kept thinking I should have more faith and that I should be able to handle this better. He basically said, "Lighten up. Infertility is a very big deal. The faith you get isn't something you are just supposed to manufacture, it's something that you get from God." I was trying to take everything on myself instead of asking God to help me with it. I did have faith, but through your life it's a constant journey or process to develop your faith further. This was another way that my faith was being developed. Just because you can't be a Zen master about this whole thing doesn't mean you have no faith.*

Rachel, another of my patients, reached a similar conclusion.

> *My relationship with God is much stronger now. There have been times that I've been angry with God, and at those times I tell him exactly how I feel. The psalmists always told God how they felt, and it was by talking to God about their anger that they would get back to praising and being in a relationship with him. He can handle whatever I'm feeling, and it's okay for me to have those feelings—we work through them together.*

Infertile women ask God, "Why are you doing this to me? Why aren't you giving me a baby?" As Barbara says, we are all on our own path. There are lots of things we don't know the whys of. But we can ask God for the faith to accept that what happens to us is happening because the path we're choosing to pursue—parenthood—is not the path that God has chosen for us. I know, that answer is not very satisfying, but it is the answer that women of faith who struggle with infertility often come to.

Eventually many reach the point where they can faithfully accept the fact that God has a plan that, although different from the plan we would have had for ourselves, will be right for us.

It's comforting for some women, to think about it this way: Instead of believing that God is ignoring their prayers, they choose to trust that God is indeed answering their prayers—but instead of the answer being "Yes, here's your baby," the answer is "Yes, I will give you what you need, but what you need is not what you think you need." The best way I can come up with to describe it is to pass along this story that a friend e-mailed to me recently:

The only survivor of a shipwreck was washed up on a small, uninhabited island. He prayed feverishly for God to rescue him, and every day he scanned the horizon for help, but none seemed forthcoming. Exhausted, he eventually managed to build a little hut out of driftwood to protect himself from the elements and to store his few possessions. But then one day, after scavenging for food, he arrived home to find his hut in flames, the smoke rolling up to the sky. The worst had happened—everything was lost. He was stunned with grief and anger. "God, how could you do this to me?" he cried. Early the next day, however, he was awakened by the sound of a ship that was approaching the island. It had come to rescue him. "How did you know I was here?" the weary man asked of his rescuers. They replied, "We saw your smoke signal."

As so many of my patients tell me after they have moved beyond infertility, through adoption, by choosing to remain childless and devoting their maternal energy to other endeavors, or by using donor gametes or donor embryos to conceive, God doesn't close a door without opening a window, as the old cliché goes. Yes, it's hard to arrive at this understanding, whether you are an intensely spiritual person or someone who is just moderately religious. But if you can get to that place, you'll find a great measure of peace of mind, as is the case with my patient Martine, who chose to adopt after several years of infertility.

The philosophy I grew up with changed profoundly with infertility. I had believed that if you work real hard at something, you get it. That was the pattern of my life. In forty years nothing bad had happened to me—I had lived a very charmed life. I didn't understand failure. But after infertility and a horrible adoption experience—the birth mother changed her mind a week after the baby was born, which utterly devastated me—followed by two successful adoptions that gave me my wonderful children, I realized that God works in some very interesting ways. Sometimes he does things that make us all a little more human. All of this certainly humanized me and made me much more empathetic about everything. It made me realize that even if you try real hard, things don't always work out the way you want them to. But they do work out the way they're supposed to. I have a lot more faith now that dark clouds have silver linings, even when you can't see them at first.

Why do you think God has chosen for you to be infertile, either permanently or for several years? In other words, when God closed the door of easy fertility, which window do you think was opened for you? As part of a research project she was working on, Barbara asked a number of infertile women this question. Here are some of the answers they gave:

- "I really needed to work on myself and my relationship with my husband. This time without a baby hasn't been wasted, because I've spent it learning more about myself and building my marriage."
- "I was sexually abused as a child, and having several years of infertility gave me a chance to come to terms with my abuse. If I'd had a baby at a younger age, I'm not completely sure I would have mothered well."
- "I think I wasn't prepared for motherhood. I've had time to learn and grow, and I'm better suited to motherhood now."
- "Giving birth wasn't my path. I realize now my path was to adopt a child and to help others adopt."
- "I have learned how to be independent. I'm not living my life for other people's approval."
- "I have learned how to value myself in ways other than being a mother. My husband has always said, 'I love you for who

you are,' but I never truly believed that. I had to learn that I'm valuable even without being a mother."

- "There are mothers in the world who have never had biological children—Mother Teresa, Mother Meera—perhaps God has in mind for me a mothering role that does not involve biological children."

I often hear from former patients years after their infertility is resolved. Looking back, they all seem to have an understanding of why they think God made them infertile. So many of them tell me that they see, with several years separating them from their infertility experience, that there was a reason for it. Eileen is one such patient.

> *After three years of infertility we chose to adopt from overseas. As I was on the plane going to pick up my baby, after all those years of infertility treatments and a miscarriage and months of red tape with the adoption agency, I realized: This is what I was meant to do. It was so clear. I was meant to be this child's mother. I really feel very spiritual about it. I couldn't be happier with my adopted child—being her mother is what I feel I was meant to do. It's a wonderful, wonderful experience.*

The passage of time gave Kerry a deeper perspective as well. After several years of infertility she has chosen to remain childless, and she derives strength from both the Jewish faith she grew up with and her more recent connection to Native American spiritual traditions.

> *I went to law school and majored in juvenile law. Now I'm working as a mediator, helping children. I don't know that I would have done that if I'd had my own children. I probably wouldn't be working at all if I had a baby. I believe this was meant to be, maybe because I can help more children this way.*

Eventually many infertile women who seek spiritual enlightenment learn to accept and surrender to—even embrace—what they think of as the will of God. They may not know why—often it becomes clear to them later on, however—but for some reason God has chosen a path that does not include biological children. "I remind myself that God has a purpose in this," says Rachel. "He's trying to teach me something. I don't know what, but I believe he knows best. God knows everything, and I know such a little piece."

In other words, thy will be done.

CHAPTER NINE

The Nitty-gritty of Infertility Treatment: Financial and Medical Issues

One of the more frustrating parts of infertility is that it turns what should be one of the most joyful, personal, intimate events of your life—the conception of your child—into a business transaction. Instead of romping around in bed with your husband, drunk with love, you're consulting with doctors, tussling with nurses, haggling with insurance companies, and constantly keeping an eye on your net worth to make sure you can even afford to do what your doctor says you need to do to have a baby.

Infertility throws you, sometimes quite suddenly, into a labyrinth of financial and medical decisions and conflicts. From the very beginning, you're faced with difficult choices, from picking a doctor to deciding on what treatments you can endure as well as afford. As painful as it may be, however, the business of infertility is unavoidable. Even if you love your doctor and have insurance that covers every last one of your treatments, the cost of infertility treatment requires that you learn to negotiate insurance benefits and manage finances carefully to optimize your options.

You can avoid some of the stress of financial and treatment decisions with careful planning, smart decision making, and coping skills. In this chapter I'm going to walk you through some of the more intractable medical and financial problems you'll face as you enter the world of infertility treatment. I'll also offer advice on what has worked for many of my patients and suggestions for how you can use mind/body techniques to relieve stress before, during, and after medical procedures. You can't avoid all potential problems, but by being prepared you can at least take away some of the stress of dealing with the unpleasant business of infertility.

PROBLEM: Should I see my OB/GYN for infertility treatment, or should I go to an infertility specialist?

ADVICE: Here's my rule of thumb: Be aggressive in seeking infertility treatment, and limit the time you spend with your OB/GYN generalist. There are many OB/GYNs out there who would argue with me on this one, but I stand firm in my opinion. Let me tell you why. Basically, there are three levels of infertility care providers: an OB/GYN, an OB/GYN who subspecializes in infertility, and a reproductive endocrinologist. While it's possible your OB/GYN keeps up with all the latest infertility news, you're much more likely to receive state-of-the-art care from a specialist. According to the American Society of Reproductive Medicine, you should see an infertility specialist if you don't conceive after six months of unprotected sex if you're over thirty-five or if you have a history of any kind of reproductive-system health problems, and after a year if you're under thirty-five. Some people stay with their OB/GYN for the first round of tests—a day-three blood test, semen analysis, a thyroid test, and the usual blood work. But don't spend too much time with an OB/GYN. I can't tell you how many of my patients have said they wish they'd seen a specialist sooner than they did. Here's what my patient Zoe says:

We started trying to get pregnant when I was thirty-one. I sensed right away that something was wrong—I think you know when something is wrong with your body, and I felt very strongly that there was a problem. I went to my OB/GYN and asked him if there was anything we could do. He said to keep trying for a year. I didn't want to do that, but he insisted. Six months, nine months went by. Finally I went to another doctor, and he ordered all the tests. It turns out my husband had a varicocele [an abnormal twisting or dilation of the vein that carries blood from the testes back to the heart] and I had a bad ovary. With problems like those, my chances of conceiving naturally were very low.

Once you've made the decision to see a specialist, I recommend, if at all possible, that you see a reproductive endocrinologist. These physicians have had extensive training in treating infertility, they keep up on all the latest research, they have the most experience, and they tend to have more in-house equipment for testing and treatment. Unfortunately, however, there are fewer than a thousand of them in the United States, so you may not live near one. If that's the case, look for an OB/GYN with a subspecialty in infertility. There are a lot of them out there. Although they may not have the ability to do extensive testing, treatment, blood work, and ultrasound in their offices, and although they lack the certification that reproductive endocrinologists have, the very best OB/GYN infertility specialists can be just about as good as reproductive endocrinologists. If you live within driving distance of a reproductive endocrinologist—some people actually drive a few hours to see one—that's the way to go. But if you don't, and if you can find a well-regarded, experienced OB/GYN infertility specialist, that should be okay, too.

There's also an emotional reason to skip the OB/GYN, and for some women this is very important: In an OB/GYN's office you're very likely to find yourself surrounded by signs of fertility—pregnant women in the waiting room, pregnancy magazines on the coffee tables, even pictures of babies on the walls. You're highly unlikely to see any of this in the office of a reproductive endocrinologist.

No matter which kind of physician you're seeing, always ask her about her experience with your particular problem or with the treatment she is offering, and if she doesn't appear to have a lot of experience, consider a different specialist. Say you have scarring from endometriosis in your pelvis and the doctor recommends removing the scarring through laparoscopic surgery. Ask, "How often do you treat endometriosis? How often do you do this kind of procedure?" In the case of laparoscopy, ask your doctor how often she does therapeutic laparoscopy versus diagnostic laparoscopy, which is much less difficult. You don't want to take a chance with someone who does therapeutic laparoscopy for endometriosis every six months—you want someone who does it twice a week. Removing scar tissue is painstaking work, and an inexperienced doctor may not do a good enough job. If possible, see a surgeon who specializes in microsurgery for endometriosis.

PROBLEM: I'm going to see a specialist for the first time, and I want to be fully prepared. What should I take along to the appointment?

ADVICE: The specialist is going to have a lot of questions—how long have you been trying to get pregnant, what birth control have you used, were you ever pregnant, what is your medical history, how regular is your midcycle intercourse, and so on. To facilitate this discussion, have the answers close at hand. If you've kept a temperature chart, take it along. If you have any test results or records from your OB/GYN, have them with you, too. Research your health insurance prior to your visit and understand what your coverage is for infertility treatment. Be able to rattle off dates, numbers, and other data. Prepare a cheat sheet for yourself, or, better yet, type up your reproductive history in a memo that you can give the doctor and that can be included in your chart. That way you'll know the doctor has accurate information.

Be honest with your doctor. It's amazing, but people often fib to their physicians in order to hide undesirable habits or to appear more health conscious than they really are. This is a huge mistake. If you've been trying for a year and your husband has been impotent part of the time, tell the doctor, even if you and your husband feel embarrassed about it. If you've been trying to conceive for a year but you've been traveling a lot and may have missed midcycle sex a few times, 'fess up. If you've had an abortion, or if you smoke, or if you don't exercise, tell the truth. All these factors can affect the course of action your doctor chooses to take.

PROBLEM: My best friend is infertile, and her treatment plan is completely different from mine. I'm worried that my doctor isn't doing the right things for me.

ADVICE: One of the worst things you can do is to compare your treatment plan with a friend's or a cousin's or a sister's or with the plan of the woman sitting next to you in the doctor's office. Everyone's body is different. We all respond differently to medication, we all have a different medical history, and we all have different financial constraints. If your doctor puts you on Clomid and your cousin's doctor refers her for IVF, it doesn't mean your doctor is an idiot. It may mean that you're not ovulating regularly and she's got blocked tubes. Or it may mean that your cousin's husband has a male factor that he's made her promise not to tell anyone about.

That said, don't just accept your doctor's advice without question. If you have concerns about your doctor's decisions, based on what you've read or what the women in your support group have said, sit down with your doctor and talk with her about it. You might even mention your friend's treatment and ask your doctor why she's not treating you in that way. There's probably a very good reason. You'll be less stressed and feel more confident in your doctor if you understand why you're being treated as you are.

PROBLEM: I'm dissatisfied with the care I'm receiving at my infertility clinic. The medical staff forgets to call with test results, they keep me waiting forever in the office, and I have to keep reminding them to schedule my test and treatment appointments.

ADVICE: This is a major complaint among my patients. Medical-office troubles are a very real, very common problem, and they cause an enormous amount of stress and anxiety. That's why I encourage you to be your own health advocate, whether you feel comfortable in that role or not. No matter how good a physician and a clinic are, they can make mistakes. I had a patient a couple of years ago who was seeing a physician I really liked. She was having advanced infertility treatment and had been seeing this doctor for a few years. As my patient and I discussed her case, it came up that she had never had the routine blood work that is supposed to be part of the initial infertility workup. Somehow she had slipped through the cracks and hadn't had the simplest tests done. After I pointed this out, she had her blood tested, and it turned out she had abnormal thyroid function. A malfunctioning thyroid can have a major impact on your ability to get pregnant, and it's so easy to correct—a pill a day solves the thyroid issue, for most people, and often the infertility, too. This poor woman went through several years of infertility because of a problem that could have been diagnosed and treated in a snap by her family doctor or OB/GYN, but somehow that test was not done. Probably when my patient saw the physician for the first time he told either her or the nurse to schedule blood work and, for whatever reason, it wasn't done. Then, the next visit, he probably assumed it had been done and the results were all negative, because nothing in her chart indicated otherwise.

How can you avoid falling through the cracks? There are a couple of things to do:

1. Accept the fact that you will have to be proactive in the care you receive and that no one will care as much as you will about your getting pregnant.
2. Educate yourself on your condition. Read books, talk with the people in your support group, look it up on the Internet—but be careful about what you read on the Internet because a lot of it is wrong. This can be painful to do, because you may find that the more you learn, the more anxious you become. But the more you know about your own diagnosis and the way in which it is typically treated, the better you'll become at being your own advocate.

3. Get a copy of your medical records once a year. Resolve recommends this, and I heartily endorse the idea. You can get a copy of your records by calling (or, in some states, writing a letter) to your doctor's office and requesting them. You may have to pay for copying costs, but the medical office is legally required to comply and provide you with copies within a designated amount of time that varies from state to state (usually a couple of weeks). When you receive your records, go over them carefully and check for accuracy and completeness. Is anything in your record wrong? Is there anything that you didn't know about? (I had a patient recently who'd had cervical cancer in the past, and during an infertility workup she was given a Pap smear. She heard nothing from the doctor's office and assumed the test was fine. Three months later she asked for a copy of her records and discovered that the Pap smear had been abnormal, but nobody had notified her of this.) If you're having problems understanding the medical lingo in your records, get in touch with your local Resolve chapter and ask if they can recommend a nurse or counselor who can go over them with you and sort through any confusion. Or sit down with your doctor or one of the nurses in the infertility office and ask him or her to explain everything. You have a right to know what's in your records.
4. Keep an infertility log. Write down everything—dates and times of tests and treatment, test results, conversations with doctors and nurses, details about periods, cramping, and all other symptoms. Make a note of your emotional states as well. That way you can look back over several weeks or months, uncover patterns of anxiety or depression, and use that information to schedule relaxation sessions, self-nurturing strategies, therapy appointments, or any other activities that shore up your emotional strength during the most difficult times.

PROBLEM: I disagree with the course of action my doctor is recommending.

ADVICE: Speak up! Your role as an infertility patient—as any kind of patient, in fact—is not to sit back and let your doctor make all the decisions. You need to be involved. If your doctor suggests a course of action you don't like, indicate that you would like to investigate other treatment plans instead. If your doctor is unwilling to discuss options with you, you may need to look for a new doctor. You definitely need to remember, however, that your doctor knows a lot more about infertility and infertility treatment than **anyone else** you've spoken to. You have to balance his knowledge base with your gut feeling. Consider the experience of my patient Betsy.

> *My mother had trouble conceiving, and I think because of that I assumed I would, too. After a year of trying, we started treatment. It was terrible. Clomid made me crazy, I had reactions to some of the drugs during my two IUIs, and I was unhappy with the hospital where I was being treated, so I switched to a different one. The doctor there started talking about doing Clomid challenges and IUIs, and I said, "No, I can't go through all that. You've got to let me do an IVF cycle. You just have to." So they said okay, and I did one IVF cycle, and that was it, I got pregnant.*

Kyra also found it necessary to make a change.

> *After I took Dr. Domar's program, I felt empowered to take control of myself, my life, my body, my doctor, and my treatment. When I was still with my previous doctor and practice, I was just going along thinking that they knew best, and that they should decide everything. But once you realize you can take control, you think, "This isn't right that they're calling all the shots. This is about me. This is about my family. I need to control this." I switched clinics, and now I tell the doctor or the nurse if I disagree with what's happening. I tell them, "I don't feel good about the*

fact that you're making me do this" or "I don't agree with the dose that you're recommending." You can change your treatment—you just have to speak up.

Once one of the nurses disagreed with me on when to do a certain procedure. I said, "Listen, I know what's going on with my body, and this isn't the right thing to do." But she didn't agree with me. So I secretly e-mailed the doctor, and she said that I was absolutely right. I felt so sneaky, but I have developed a real sense of entitlement to control my treatment. I would never ever in a gazillion years have done that before.

PROBLEM: My doctor is not very good at communicating. Sometimes when I leave his office after an important meeting, I have more questions than I had at the beginning of the appointment! Then, when I call him to ask follow-up questions, he takes forever to return my calls and is gruff and impatient on the phone.

ADVICE: An appointment with the doctor can be a dizzying event, particularly if it's a consultation with a new doctor or a meeting in which you'll be going over the details of a long-term treatment plan. You may feel anxious and worried and full of questions.

You can get the most out of a doctor's appointment or consultation when you're as calm and clearheaded as possible. I recommend doing a twenty- to thirty-minute relaxation just before you go to a very important appointment, followed by as many minis as you can manage while you drive to the office, sit in the waiting room, and meet with the doctor. Relaxation will help you stay calm and collected, ease tension, and keep you from panicking if the doctor says something upsetting.

No matter how calm you are, however, it's best to bring someone who can take notes. If your husband can do this, great, but if both of you are anxious, consider asking a friend or family member to come along and take notes while you and your husband talk with the doctor. Research has shown that when you're very anxious before a doctor's appointment, as much as 90 percent of what the doctor says goes right over your head. So have someone there to take everything down on paper. Toward the end of the appointment summarize what the doctor said back to him. For example: "So what you're saying is, my age is a concern, but my blood levels are within normal range, and what you recommend is three cycles of injectible medication with IUI, and then we'll go on to IVF if necessary." By repeating the information back to him, not only do you reinforce it in your own mind but you confirm that what you heard is what he said.

Be especially careful to delineate what steps need to be taken and who needs to take them. Will you do the scheduling legwork, or will the nurse or doctor take care of that? This is really important—I have seen patients lose months of valuable time because they think the doctor is going to call to schedule a procedure, and the doctor thinks the patient is going to call.

When your doctor is discussing treatment, be sure to ask five things:

1. What procedures are you recommending?
2. What time frame do you have in mind?
3. What are the chances of success for this particular treatment plan?
4. What are the alternatives to this plan?
5. What will we do if this plan fails—in other words, what is Plan B, and when will we go to it?

As far as questions go, ask him how and when you can contact him with follow-up questions, because you just *know* you're going to think of a million questions the minute you leave the office. If he says, "Call me anytime," but you know it's hard to get through to him, ask for options. Can you call him early in the morning or late in the afternoon? Can you e-mail or fax questions to him? Can you schedule a short follow-up appointment? Try to pin him down.

Over the next couple of days, do some research. Talk with your support group about the proposed treatment plan. As you consider it, collect all your questions on a list. Then, when you talk with your doctor, you'll have them all right in front of you, on paper.

PROBLEM: My doctor is always in a rush, and I barely have a chance to talk to her.

ADVICE: You have a right to your doctor's time. You're paying a lot of money, perhaps out of pocket. You need to understand what your doctor is doing or saying—it's too important to get this stuff wrong. If she's in a hurry to get out of the examining room and on to the next patient, tell her that you need her to slow down. Say something like "Doctor, you seem to be in a rush, but I need you to answer my questions. I need to make sure I understand what the plan is, what I need to do, and what your next step is. If you don't have time to talk with me now, please tell me when you will have time." If the problem persists, you may need to think about getting a new doctor, as my patient Nancy did.

> *Early on in this whole process, one of my doctors made a mistake—she didn't order a day-three test, which is a standard part of the first-round testing. After that happened, I realized I had to be in control. I had to take every piece of control that I could. Otherwise I couldn't deal with it. So I get really involved in talking to my doctors and the nurse coordinators, and when something doesn't work, I need answers to my questions. When I met with my second doctor, I told her this. She said she was fine with it, but then she would be completely unavailable to me. Also, one of the nurses was very rude to me on the phone. And another told me the wrong thing regarding when my next procedure would be. Five minutes later the doctor called back and said the nurse had made a mistake, and she rescheduled the procedure. And I said to the doctor, "You know what? I can't have this. I can't have nurses making mistakes and you being impossible to reach." Also, I never felt that she spent time really thinking about my specific situation. When she called me, I could tell she was reading my chart as she was talking to me on the phone. I said to her, "Did you read that before you called me?" and she said, "No, but I'm reading it now." And I said, "Please, the least you could do is read it first before you call me."*

So I switched doctors again, and now I'm happy with my new doctor. He really listens to me and takes time to answer my questions. It's such a nice change. One of the things I told my new doctor was that I needed to know about everything that was out there in terms of treating infertility. I wanted to know every single resource available to me. And he did a great job with that.

PROBLEM: I just don't trust my doctor. He is an expert in the field, but I just don't feel good about him.

ADVICE: Don't discount the importance of having good chemistry with your doctor. The doctor-patient relationship is like any other—you have to respect, trust, and feel comfortable being around him, or you may not be satisfied with his recommendations. You may be seeing a doctor that everyone in town says is fantastic, but if you can't stand him, you need to find someone else. You have to be able to trust your doctor, or your anxiety will multiply, as it did for my patient Sophie:

We met with the two infertility doctors in our town, and we didn't like either one of them. The first didn't have a great reputation, and I didn't feel good about that. The second was very nice, and I liked him very well, but his advice seemed a little out of date. So we decided to go to a big hospital in another city and meet with a doctor who is nationally known for his work. I liked him personally, and I had lots of confidence in him. I don't think I would have trusted either of the first two. Yes, it was a lot of back-and-forth going to the other city, but it was worth it.

You also have to know how much chaos you're capable of taking. One of the best infertility programs in the country is, unfortunately, one of the least organized. This is a shame, because the doctors in this program are excellent at treating infertility. But overall the program is a pain in the neck for patients. If your clinic is like that, you have to ask yourself what your chaos quotient is. For patients who are highly motivated and don't mind monitoring every move their medical team makes, a program like this can work. But if the lack of organization drives you crazy, find another program where you'll feel more comfortable.

If you don't like your doctor, maybe you can see her partner—patients switch from partner to partner all the time, and doctors usually don't take it personally. Who knows? Maybe if you can't stand the doctor, she can't stand you either.

Your relationship with your doctor is like any other relationship in your life: In order for it to work, you need the right chemistry. Many times a patient in a group will complain bitterly about the bedside manner of an infertility specialist whom another patient in the same group absolutely loves. Your relationship with your doctor is very important. You need to feel that you can communicate with and trust him. Just because your second cousin loves a certain doctor and conceived twins does not mean that doctor is right for you.

Also, try to keep in mind that your doctor is human and has bad days sometimes, just as you do. If you have a meeting with your previously reasonable doctor and he seems irritable, cuts you off midsentence, and dismisses your questions, remember that in all likelihood, the problem does not lie with you. He may have had a fight with his wife that morning, he may not be feeling well, or maybe one of his favorite patients just miscarried. If this behavior continues, go elsewhere, but all of us have bad days once in a while. Doctors are no different.

I know that this is the least of your concerns, but doctors get frustrated when treatment fails. A doctor can do a six-hour microscopic laparoscopy and remove every visible dot of endometriosis, but if you don't get pregnant, his surgery is a failure. He can coordinate a perfect IVF cycle, administer the exact dosage of meds, and skillfully transfer beautiful embryos, but that skill doesn't guarantee a pregnancy. Doctors get attached to patients and desperately want treatment to succeed. When you sit in a doctor's office and cry, sometimes it can make him feel that he's failing you. I remember so clearly the look on my doctor's face during the ultrasound that showed that I had just miscarried. He looked crushed. As I cried, he silently handed me tissue after tissue. I later told him that I knew how hard it must have been for him, delivering such sad news. He said that it is by far the worst part of his job.

If you lack trust in your doctor, seek a second opinion. I heartily endorse second opinions in infertility, even if you adore your doctor. When is a good time to have a second opinion? After an initial consultation when a doctor designs a new treatment strategy, if you've been doing the same treatments several times and you're getting nowhere, if your doctor recommends something you don't feel ready for, or even if you just have a nagging feeling in your gut that something is wrong with the way your treatment is going.

Schedule a second opinion by calling another doctor and requesting a second-opinion appointment. Get a copy of your medical records and treatment strategy and take them to the new doctor. Be sure to take along any records, schedules, or test results that she may need to see. If the second doctor says that what you're doing is totally sound, you know your doctor is on the right track, and you can feel confident. If the second doctor has a different opinion, though, you then have to decide which doctor to go with—or seek a third opinion.

Be sure to tell your doctor that you're seeking a second opinion. Don't worry that your doctor will hold the second opinion against you. Good doctors don't mind second opinions—they're quite common in infertility—and most doctors actually welcome them. Some do take it personally—the younger ones, especially—but they shouldn't.

I'm not recommending doctor shopping, however. Some infertile women go from doctor to doctor to doctor in the desperate hope that one will offer a guarantee that the others won't. I've had a lot of patients who doctor shop in a bad way. For example, I recently saw a forty-two-year-old woman with a high FSH level, which means her chances of conceiving with her own eggs are not good. She went to see Dr. A, who said that because of her age and FSH level, it's very unlikely that she'll respond to IVF. He recommended egg donation. Dr. B said the same thing. But Dr. C said,

"Oh, I can get you pregnant, no problem, just sign on the dotted line and give us twelve thousand dollars." This guy was most likely a charlatan. Unfortunately, there are doctors in all regions of the country who will treat you despite a very poor prognosis. You have to be careful. There's a lot of money to be made in the field of infertility, and, sadly, it attracts some dishonest doctors.

But wait—how do you know that Dr. C is a charlatan and not a brilliant doctor who knows something the other two don't know? That's tricky. You can try getting in touch with your local Resolve chapter to see what they know about Dr. C. Ask friends, the women in your support group, even your family doctor or OB/GYN or any friends you may have who are physicians or nurses. If you hear a lot of terrible stories, Dr. C is probably a fraud. Knowing as much as possible about your diagnosis helps here, too. If you're well educated, you'll have a sense of which doctor is being honest with you. And remember the old adage: If it sounds too good to be true, it probably isn't true.

Eventually you have to let go and trust your doctor. That doesn't mean you should ignore mistakes, accept mediocrity, or stop educating yourself. Nor should you take everything your doctor says at face value. You do have to let her do what she was trained to do, however.

PROBLEM: My doctor did something that made me very angry. What should I do?

ADVICE: There are three good ways to deal with a conflict with a doctor:

1. Talk to him face-to-face. Make an appointment, sit down with the doctor (after doing a nice long relaxation and loads of minis), and tell him why you're upset. Write out a list of talking points ahead of time so you can be clear and to the point. Ask him to explain why he did what he did and what he's willing to do to remedy the situation.

2. Write a letter. Face-to-face conflict resolution can be very tough, particularly if confrontation makes you nervous or if you feel intimidated by people in authority. If you dread the thought of an in-person meeting, write a letter. Fully explain your concerns, and tell your doctor what you would like him to do to make amends. End the letter with something like this: "I have to know whether your team is capable of providing the comprehensive care I require, and if not, I need to find a different doctor."
3. Contact the medical board. If you think your doctor has done something truly illegal, improper, or unethical, write or call the medical board in your state to make a formal complaint.

PROBLEM: I get terribly anxious before doctor's appointments, tests, and procedures.
ADVICE: Dig deep into your infertility coping-skills toolbox, because there are many ways to relax before, during, and after encounters with medical professionals. Here are some examples, and be sure to refer to Chapter 2 if you need to brush up on your technique.

- **Mini-relaxations.** I cannot tell you how valuable minis are before you see a doctor, talk to a nurse, as you go through a treatment, as the technician inserts a needle for a blood test or probes during an ultrasound, when you place or receive a phone call from a doctor or nurse. Minis calm you down and clarify your thinking. They can also take your mind off nervousness and pain, as my patient Bobbie found.

The first time I gave myself a shot, I was petrified! My palms were sweating like crazy, and I was thinking, "It's going to hurt." The needle looked enormous, and I kept thinking that if I waited a little longer maybe it wouldn't hurt as much. I got myself prepared to do it twenty times, but I couldn't do it. So I did a mini, and I found myself getting a grip. Then I gave myself the shot, and it wasn't so bad. After that I always did minis to calm down before shots, although after a while I didn't need to. Eventually it was a breeze—one time I gave myself a shot in the bathroom at a black-tie dinner.

- **Longer relaxation.** Continue to spend twenty to thirty minutes a day eliciting the relaxation response. If you relax regularly, the beneficial carryover effect lasts throughout the day. You can also elicit the relaxation response during a medical procedure, as many of my patients do. Bring along your relaxation tapes if you use them. Notify the doctor, nurse, or technician that it's important for you to relax and that you'd prefer it if he or she talked to you only when necessary. If, like my patient Gwen, you relax on a regular basis, you'll probably be able to move relatively easily into a more relaxed state.

Before I go for an ultrasound to see how many follicles are developing, I take some quiet time just to sit and be peaceful. I remind myself that I am not in charge, and that whatever is on the screen is the best my body and I can do today. I can't ask more of my body than whatever it's doing. Relaxation helps me to remember that it's not worth it to beat myself up or torment myself. I've done that, and I'm sure everyone else has, too. But when I slow down and relax, I can remind myself that my body and I are doing the best we can.

- **Guided imagery.** Taking an imaginary vacation to a truly special, peaceful place can help take you away from the obtrusive sights and sounds of a medical office. Close your eyes, travel in your mind to a sandy beach or a woodsy forest or wherever you feel at peace, and imagine the calming sights and sounds of your getaway space. If it helps, listen to a tape of relaxing music or soothing nature sounds.

- **Cognitive restructuring.** This is crucial when you've just had a consultation and the doctor said one negative thing and several positive things. He might have said, "I'm concerned with your age, but your FSH level is good, your tests are all negative, and your husband's sperm count and motility are just fine." If you're like most women, you'll forget all the positive things the doctor said and fixate on the negatives, berating yourself with a constant stream of "I'm too old to have a baby, I'm too old to have a baby, I'm too old to have a baby." Put it in perspective through cognitive restructuring.
- **Self-nurturance.** I recommend that you nurture yourself daily, but I urge you to go for an extra-big dollop of self-nurturance after run-ins with the medical establishment. Take yourself out to lunch after appointments. Treat yourself to a new pair of shoes after a blood test. Stop at the botanical garden for a walk on the way home from an ultrasound. If you're rushing back to work after your appointment, play your favorite music in the car along the way, or try to make time later in the day to nurture yourself.
- **Yoga.** A yoga session after your appointment can do wonders to help you relax and to give you a feeling of control and connection with your body. Medical procedures can often make infertile women feel alienated from their bodies; yoga can re-create a bond between mind and body.
- **Group support.** Lean on the fellow members of your group, either when you meet with them in person or over the phone. Share the details of your appointment, and ask for feedback. They may have great advice on questions to ask your doctor, tips on handling that particular procedure next time, or suggestions on coping with any side effects—for instance, cramping from an

IVF retrieval—that may occur. They'll probably also offer a healthy dose of empathy, which is just what you need after a hard time with the doctor!

PROBLEM: My doctor is fine, but the staff members at my clinic are very impersonal. It's a big clinic, and I feel like I'm getting lost in the shuffle.

ADVICE: Find one nurse you can establish a relationship with. Call her by name, be friendly to her, and treat her like a professional. A really good nurse is a fantastic source of information, both about treatment and about how to deal with the doctor and the other clinic staff. Cultivate the relationship by going out of your way to say hello to her each visit, asking her about her family, whatever. (An occasional box of chocolates for the nurses or receptionists doesn't hurt either.) You may even find you can ask her questions and get answers a lot faster than you do from your doctor.

Let me give you an enormously important piece of advice here: Be nice to the nurses! They have a tough job, being sandwiched between very anxious patients and very harried physicians. They do a lot of the clinic's dirty work—for example, in many infertility offices nurses call patients with negative pregnancy-test results, but doctors call with positives. That means the nurses never have the joy of calling with good news. Infertility nurses tend to have high turnover, are really stressed out, and get blamed a lot by patients and doctors. Being nice to the nurses can only help you. Yelling at the nurses will get you absolutely nowhere.

Many of my patients complain about rude nurses and receptionists. Margaret, a bubbly, personable patient of mine who lives in a rural area and had to travel a long distance to her infertility clinic for treatment of her endometriosis, found the doctors and nurses very impersonal, and that was hard for her to take.

I've been to so many doctors over the years where there is no sensitivity going on. It was terrible in the clinic where I had my most recent treatment. I was there every other day, and nobody learned my name. Every time it was like they'd never seen me before. When I got my period, nobody said "I'm sorry" or anything. I think they didn't want to feel responsible for the fact that I hadn't gotten pregnant, and they acted that way because they didn't know what to do. They continued to treat me very impersonally until I tested positive, and then they were all hugs and congratulations.

If one of the staff members at your clinic says something bitchy, try to react calmly. (A mini comes in handy here.) If you can think on your feet, say something like "How do you suppose that makes me feel?" or "Why did you say such a hurtful thing?" Otherwise go home and be upset, and when you calm down, think about giving her a call. Say, "I really want to work with you, and we both have the same goal of getting me pregnant. But your comment hurt me. Is there a way that we can move beyond this?" It's best if you're able to work it out—maybe she was just ultrastressed and took it out on you. She may apologize and become your friend. (If so, teach her how to do minis so she doesn't keep blowing up at patients.) If that doesn't work, or if the nurse did or said something truly inappropriate or offensive, take it up with the doctor or office manager, or leave the practice, as Ellen did.

We had a terrible experience at our first clinic. My husband is a well-known local sports figure. I had a miscarriage, and during the D&C one of the technicians recognized me as his wife. He proceeded to tell me that many of the people on the hospital staff—including the nurses in the room—were fans of my husband's. It was a real confidentiality issue. I could imagine everyone going home that night telling their friends and family, "Guess whose wife had a miscarriage today?" I switched clinics right after that.

PROBLEM: Our health insurance doesn't cover infertility treatment.

ADVICE: This is such a huge problem that I'm going to spend the rest of the chapter on it. As of this writing, fewer than a dozen states mandate complete coverage for infertility treatment—don't even get me *started* on how unfair that is. We need to find ways to cope with the financial woes of infertility. Here are some of the suggestions I give my patients:

Know your own coverage. The majority of states don't require health insurers to cover all aspects of infertility treatment. But here's something that many of my patients don't realize: Many insurance policies do cover *some* portion of treatment, even if they don't cover everything. That's why it's crucial for you to find out *for sure* what your policy will cover. Get it in writing, and don't go by what the human resources rep at your job says. She may not know the intricate details of the company's health insurance policy—I've seen this happen many times—and she may say there's no infertility coverage when in fact there is. In some cases, for example, IVF isn't covered, but some of the diagnostic or therapeutic procedures leading up to it are. Someone who is ignorant of infertility, such as your HR rep, might make a mistake on this fine but costly detail. I've heard many stories about HR reps making mistakes—for example, one told a patient of mine that even though the patient lived and worked in Massachusetts, where infertility coverage is mandated, her treatment wouldn't be covered because the company is based in a state where it is not. And that was completely wrong. Get a copy of the actual policy and find out exactly what is and isn't covered. Your HR department should have the policy on file—and when you look at it, make sure it's not an old, outdated version. Make a copy, keep it in your files, and ask for a new copy at least once a year or whenever your company revises or amends the policy.

Even if the human resources representative (and her supervisor and her supervisor's supervisor) tells you that there is no insurance coverage for infertility in your policy, read it carefully anyway. And speak to your physician about the situation, because it's possible that the way she reports your treatment or diagnosis can affect whether insurance covers it. For example, if you're experiencing severe menstrual cramps and your doctor decides to do a diagnostic laparoscopy, she can state that the reason for the procedure is to rule out endometriosis, which is usually a covered service, rather than to treat infertility. If you have irregular menstrual cycles and are taking medication to induce ovulation, this can be billed as treatment for anovulation rather than infertility. Do not commit insurance fraud, but see what can be fairly billed.

Fight your insurance company. I've seen several cases where women who were initially denied coverage appealed to their insurance company and eventually won some coverage. It's worth a try.

Know your state's laws. Contact the state insurance commission to find out what, if any, state-mandated coverage exists. At last count, eleven states offered some kind of mandated coverage, but that changes all the time. You really want to do your homework here and get actual documentation—don't just go by what you've heard from other people or by what an uninformed receptionist tells you on the phone. I've had literally hundreds of patients tell me that they believed they had no infertility coverage, and when they pushed and read and investigated, they discovered they were entitled to some coverage after all. Don't believe anything you're told; ask to see it in writing. Two other great resources for information on state-by-state infertility coverage are the American Society for Reproductive Medicine's Web site (www.asrm.org) and Resolve, which has published a book on the topic called *Infertility Insurance Advisor—An Insurance Counseling Program for Infertile Couples*.

Ask for a discount. Some infertility clinics offer discounts for un- or underinsured patients. But they may not advertise it, so be sure to ask what the clinic can do for you. And there may be other money-saving options as well: In some clinics patients who get pregnant donate their leftover medications to uninsured patients—one patient of mine got three thousand dollars' worth of free Pergonal that way. (Keep this in mind if you get pregnant—don't flush those leftover meds!) Just ask your doctor, the office business manager, or that nurse you've made friends with—you never know what might be available.

Mix money and medicine. Be sure to make your finances a part of the discussion with your doctor when she's designing a treatment plan. If you don't have complete coverage, tell her exactly what your insurance will and will not pay for. If your coverage is limited—some insurers will cover up to twenty-five thousand dollars, for example, but not a penny more—make sure your doctor knows this. If paying out of pocket is a hardship, tell her that money is a real issue and that, if possible, you'd like her to design a treatment plan that makes the most of what is covered. Ask the doctor what treatment will give you the most for your money.

There are ways to design effective treatment plans while negotiating financial obstacles. Here's one example: Say you have only one open fallopian tube, or your husband's sperm count is low. Even though an IVF cycle may cost about three times what a medicated IUI cycle costs, the increased chance of success may make it worth the money. And there are other ways of stretching your infertility dollar. For example, if you hope to have more than one child, you can freeze excess IVF embryos for future use—a fresh IVF cycle can cost up to six times more than a "thaw" cycle. Likewise, if you use a donor embryo, opting for a frozen one may cost less than a fresh one. Some centers even offer a free IVF cycle to women who donate eggs. Of course, donating eggs or embryos can pose a difficult choice. You have to know your limits.

Consider dramatic alternatives. Sometimes big changes can shift your financial options. Would changing jobs give you better insurance coverage that would pay for infertility treatment? Or moving to a neighboring state? I know this sounds extreme, but sometimes an extreme change is worth it for financial coverage. I practice in Massachusetts, which has some of the best infertility coverage in the nation. Nearby New Hampshire has far less comprehensive coverage. Because of the difference, a number of my patients from New Hampshire actually move and become residents of Massachusetts just to be entitled to that generous coverage. It sounds incredible, but let me tell you this: Moving may be a lot less expensive than uncovered infertility treatment.

Watch out for "guaranteed" programs. Some treatment programs offer money-back guarantees: If you don't get pregnant after five IVF cycles, for example, they'll refund your money. Sounds good, but this really can be a flawed approach. First of all, some of these programs tend to be very selective regarding whom they accept. They prefer the easiest-to-treat women, who are under thirty-five and whose husbands have no male factor. Second, drugs are usually not included, and they can cost thousands. Third, IVF is a real grind, and not many women are game to endure five of them. With most guaranteed programs, if you don't do all five IVFs, you won't get your money back. If you do decide to sign on, make sure to have a reputable lawyer check out the contract, and keep in mind that most medical societies consider this a controversial practice.

Be aware of what financial problems can do to a marriage. Disagreements over money have the power to really test a marriage. Frequently one spouse wants to spend more on infertility treatment (usually, but not always, the wife) than the other. Sometimes one spouse wants to set aside money for adoption, and the other wants to spend everything in pursuit of a biological child, because he or she is not interested in adopting. (Remember, if adoption is your Plan B, you may need to start considering the ways to save money for it.) I can't tell you how many times I've met with couples in which the wife is willing to go into debt for infertility treatment and the husband isn't. If the wife gives in on this disagreement, she may harbor anger and resentment toward her husband for the rest of her life. If the husband gives in, he may blame his wife indefinitely for putting their financial future in jeopardy, particularly if treatment fails and they don't conceive a baby. Phoebe's experience illustrates how crucial it is to come to terms with how expensive this can be.

Infertility has been an enormous financial drain on us. We have probably spent, conservatively, fifty thousand dollars on treatment. Spending the money was very hard on my husband. He had been married before and is older than I am. We cashed in stocks and used some retirement money and other savings. This was very painful to my husband, because he was a saver. And I said, "Honey, if we're going to have kids, we're going to have to get used to spending money." We had some real battles. He eventually agreed, but it was hard. We went through so much money in such a short amount of time that it made our heads spin.

In the best situations the couple agrees on how much to spend and then stops pursuing treatment when they reach that point without achieving a successful pregnancy. But I rarely see this happen, and the result is anger, resentment, blame, and bitter disagreement. If your marriage is suffering because of arguments about money, I urge you to talk with a marriage counselor who can help you find a compromise you can both live with. (For counselors in your area, contact Resolve or the American Society for Reproductive Medicine, or ask your infertility doctor for a referral.) Many health insurance policies will pay for at least a few visits to a marital therapist; if yours doesn't, the out-of-pocket cost of a couple of appointments may be well worth spending. (If your insurance doesn't cover therapy, ask the therapist if she'll give you an out-of-pocket discount.) And if therapy is out of the question, consider asking a clergyperson to help you negotiate.

See a financial counselor. Paying for infertility treatment out of your own pocket can drain a savings account pretty quickly. How much can you afford to spend? Should you raid your retirement fund or take out a second mortgage on your house? Can you afford to leave your job temporarily while you pursue treatment? If you pay for IVF, will you have money remaining to finance an adoption? These are not easy questions to answer, but a financial counselor can help. He or she will help you figure what you can and can't afford and whether it makes sense to take on debt or borrow from retirement savings.

Be ready for unsupportive comments. If you spend thousands of your own money and tell other people about it, expect criticism and disbelief. First of all, people who don't know much about infertility are shocked to hear what treatment costs. Second, people who don't share your intense desire for a baby may not understand or agree with your willingness to part with such hefty sums. Prepare for comments such as "Why would you possibly spend so much?" or "Why don't you just adopt?" Caroline's friend was even less tactful.

> *We spent eleven thousand dollars for an IVF, and it failed. I had a friend who said, "What a waste! For eleven thousand dollars you could have toured Europe for the summer. You should ask for a refund!" I was so angry. I don't think of it as wasted money. It's a medical treatment, and medical treatments don't always work. I wouldn't ask for a refund if I had cancer and it wasn't cured.*

I recommend that you prepare responses ahead of time so you can be ready for comments like these. For instance, "Well, I guess we have different priorities than you would in this situation" often works, as does "Having a baby is really important to us right now, and it's worth any price to us."

Think carefully about accepting help from family. Sometimes a couple's parents offer to lend or give money to finance infertility treatment. Your first reaction may be to say yes immediately, but before you do, think about the possible complications that can arise, and be sure you can live with them. For example, perhaps your parents are willing or able to chip in and his aren't—will that create resentment between the families? Will your family feel more of a sense of entitlement to the baby, or will your husband's family feel less connected to their grandchild? Will your parents feel they have a right to share their opinions and criticize you when you make treatment choices with which they disagree? Will they become overly involved in your treatment program? Will they ask for a say in decision making regarding which treatments to pursue? Will they push for treatments with which you are uncomfortable, such as using donor sperm? (Using donor sperm tends to be far more palatable to the wife's family than to the husband's.) How will your parents feel if they give you thousands for a treatment and it fails? If you do choose to accept their money, will it be a loan or a gift? How will you pay it back? What if you *can't* pay it back—for instance, if you intend to repay with money from your job but then get pregnant with triplets and have to go on disability because you're on bed rest for six months? And what if your sister discovers two years from now that *she's* infertile—will Mom and Dad have enough left over to help her out, too?

If you take their money, be certain you're all on the same page about whether and when you'll pay it back (a written agreement helps) and how you'll be using it. And make sure they know that success is not guaranteed—an IVF, for example, succeeds only 25 percent of the time. I'm not saying don't accept money from family, but you should be aware of all the potential complications.

Ask your employer for financial aid. Go to both your employer and your husband's and ask if they can help out. Some companies whose health insurance doesn't cover infertility treatment will buy a rider for their policy that adds infertility coverage—perhaps the policy doesn't cover infertility because none of the employees has ever asked for coverage. If the owners of the company are trying to create a family-friendly place to work, they may go for it. If you know several other infertile employees, approach the company together and request coverage. It's worth a try. And even if the company doesn't change the insurance coverage, it's possible the owners will agree to pay out a onetime cash benefit to infertile employees.

Look into a pretax medical account. In some places you can set aside pretax money for medical care. And under certain conditions medical treatment is deductible. Check with a tax expert on this. Either way, save copies of all of your receipts, invoices, and other paperwork.

Lobby for change. As I've mentioned, Massachusetts has excellent infertility coverage, and I'll tell you why: because the people who started Resolve live here. They lobbied the state legislators like crazy and managed to get mandated infertility coverage on the books. It wasn't easy, but they did it. It makes me so angry that every state in the country doesn't have coverage for infertility. But if enough people push their legislators, we may be able to change that.

"We've spent at least thirty thousand dollars on treatment, and we still don't have a baby," says my patient Claudia, thirty-five, whose insurance covers only some infertility treatment. During the three years she's been trying to conceive, Claudia has tried Clomid, laser surgery in her fallopian tubes, three IUIs, and three IVFs. Claudia conceived after the second IVF but miscarried ten days later. The third IVF was canceled because of poor egg quality. "You just sacrifice to pay for these things—there's no new siding on the house, the bathroom is still 1970 blue. When you go out in the yard, you remember that the new patio isn't there because you did an in vitro cycle instead.

"Did spending all that money cause stress? Yes. We knew we were going to spend it, but actually letting go of it was really hard. Now we're taking a break for a while and saving for the next procedure.

"I think it's a sin that my insurance doesn't cover these treatments. I have a physical problem—a disease. In this world today a drug addict can get needles for free, but I can't. It's unfair. I think one day this will all change, but it will be a slow process. I don't have time to wait. I just want a baby."

CHAPTER TEN

When Miracles Don't Happen: Coping When Treatment Fails

Infertility equals failure—repeated, heartbreaking failure. You fail to conceive month after month after month. You fail pregnancy tests. You fail treatment. Coping with that sense of failure is one of the hardest parts of infertility, as my patient Theresa knows. Theresa, thirty-five, began trying to conceive five years ago. She got pregnant within a year, but miscarried. Six months later she miscarried again. For a long time treatment failed to help her—nine cycles of Clomid and a medicated IUI all failed.

> *You feel sorry for yourself, you feel hurt, you look at your spouse and say, "Is it you? Is it me?" And then, when you find out, like I did, that it was my fault, you feel guilty, you beat yourself up. You wish it would be him so you wouldn't have to keep blaming yourself. It just sucks.*

Finally Theresa became pregnant with twins after her second IUI, but even then her joy mixed with sadness when one of the babies miscarried. After a tumultuous pregnancy she delivered a baby girl and immediately began trying to get pregnant again. Another miscarriage followed. And another, caused by an ectopic pregnancy.

The first time I had a miscarriage, I thought the whole world was going to end. I couldn't believe it had happened to me. When you find out you're pregnant, you go to the doctor's office, and they do a blood test to confirm the pregnancy, and everyone's happy. But they don't tell you how many pregnancies end in miscarriage. When it happens, you wonder, "What did I do wrong? Did I drink a glass of wine before I knew I was pregnant?" And when the doctors told me I hadn't done anything wrong, I felt totally powerless.

I could not go into each procedure thinking this was going to be the time it would work. I had to go into each procedure thinking it would probably not happen that time. I somehow had to expect the worst while hoping for the best.

Theresa's right. As a woman being treated for infertility, you somehow have to "expect the worst while hoping for the best." And that can be incredibly difficult. When you undergo infertility treatment, you have to psyche yourself up so much to endure the medications, to live with the side effects of hormone treatment, to bear the constant blood tests and ultrasounds and injections. You have to get excited about it, because that's the only way you can push yourself to slog through the process. The problem is, when you psyche yourself up like that, when you put so much time and energy into doing a high-tech cycle, you also set yourself up for a huge crash if it fails.

Most of us tend to expect success, particularly when we work hard—it's the American way. Not only that, most women grow up thinking that when we want to have a baby, we will. The feeling of expectation grows when you see a doctor for help, particularly if you see an infertility specialist who is well known for successfully treating couples with stubborn infertility. You're sent to the big-time, well-known reproductive endocrinologist, and you think that he'll get you pregnant—you're always seeing him interviewed in the newspaper in reference to miracle babies. He does a workup, provides treatment, and of course that treatment should work. We trust modern medicine. We trust our doctors to fix what's wrong. After all, the doctor is optimistic. He said he's optimistic. And there are so many different treatments to try.

The media don't help at all. You don't pick up *Time* or *Newsweek* or the *New York Times* and read about IVF failures. You read about miracle babies, about a forty-three-year-old getting pregnant, or a 51-year-old giving birth to triplets, or a grandmother carrying twins for her daughter who doesn't have a uterus. What you don't read about is the failures. And believe me, there are a lot more failures than success stories. The vast majority of people who undergo a treatment cycle don't get pregnant. For example, an IVF cycle has about a 25 percent chance of success—that means three out of four cycles fail. A medicated IUI cycle has about a 15 percent chance of working. Every cycle you go into, the odds are against you. And yet you have to hope.

Research shows that women having infertility treatment cycles dramatically exaggerate their chance of success as they perceive it in their own minds. Even if physicians clearly delineate their likelihood of getting pregnant, most women estimate their chances as being much, much higher. Although intellectually they may understand that these treatments have a high failure rate, emotionally they don't. Their tendency is to think positively: *Of course I'm going to get pregnant. Of course it's going to work for me. Those low numbers are for other people. I'm being such a good patient; I'm seeing a really good doctor. I've been really good—I'm not drinking any alcohol, I've eliminated caffeine, I'm not exercising, I'm healthy. I'm a good person, I go to church every week—of course it's going to work for me. It has to! I'm working so hard on it.*

And then, when a treatment fails, you're plunged into anxiety, depression, and stress. You tried so hard, but still you failed. It just doesn't add up, it doesn't seem fair. Failure is even more stressful if you're paying out of pocket and you have only enough money to do a certain amount of treatment. When that treatment fails, not only are you not pregnant, you're out five thousand or ten thousand or fifteen thousand dollars or more.

The higher the level of treatment, the harder it is when it fails. Not only are high-tech treatments such as IVF intensive in terms of time, emotion, energy, and cost but they also represent arriving at the end of the line. When a woman starts treatment and her first Clomid cycle doesn't work, in the back of her mind she knows there are many other treatment options—more Clomid cycles, IUI, medicated IUI, and IVF. But getting to IVF feels enormous, because it may be the last option for you and your husband to achieve genetic parenthood. And if that first IVF cycle doesn't work, many women truly begin to face the possibility that nothing will help them get pregnant. The first unsuccessful IVF is one of the worst failures of all.

 Coping with Failure

When it comes to failure, Aileen has been there.

After six months of trying to get pregnant naturally, my doctor gave me a blood test and discovered that my FSH was twenty-five. I didn't know what that meant, so I asked her, and she said it meant I would probably never get pregnant. What an awful feeling. I had such an immense sense of failure. I felt like I had failed my husband because he had married someone older who couldn't have children. He could have married someone younger who could have given him a baby. He's the only male in his family, and it was really hard for him, but he said he was okay with it.

For me the failure was the hardest part. It was so tough for me to actually believe that it was okay with him, that he could accept me, that he didn't want to leave me, that he wasn't harboring some huge resentment toward me. It was difficult for him to get me to believe that, and that was really frustrating to him.

There's no way to take the sting out of an infertility diagnosis, a negative pregnancy test, the arrival of an unwanted period, or a treatment failure. But you can protect yourself by reducing stress and anxiety as much as possible before and after bad news comes.

Before you even begin your treatment cycle, think about how many people you want to tell about it. If thirty-five people know, be ready to receive calls from thirty-five people on the evening of day twenty-eight. If you're not pregnant, it's very hard to share this news, especially when you've known it only it a couple of hours yourself. Consider appointing a spokesperson—a sister, a friend, your mother—to spread the news so you won't have to. And remember, as you decide whether to tell friends and family that you're beginning a treatment cycle, be sure to balance the need for privacy with the need for social support. Having loving support during times of failure can be a huge help.

Plan downtime for a day or two after you expect your news. If you're doing a treatment cycle, you have a rough idea of when you'll be having your pregnancy test or when you're expecting your period. Be careful about what you schedule in your life around the time you're going to find out if you're pregnant. For example, don't agree to attend a barbecue at the house of a friend with children. And if you possibly can, try to avoid difficult work projects or out-of-town trips. I advise my patients who are doing an IVF cycle to make plans to have dinner alone with their husbands on the day of their pregnancy test. If the news is good and you are pregnant, you can celebrate privately with nonalcoholic champagne. If you aren't pregnant, you can have some quiet time together to mourn your loss and, if you're up to it, to talk about what step you'll take next.

Speaking of next steps, I always tell my patients that the best thing they can do when they go in for a treatment cycle is to have a Plan B. You don't have to stick with it, mind you, but you should have something to hold on to if your treatment fails. For example, if you're on your first Clomid cycle and it doesn't work, Plan B can be to do another Clomid cycle. If Clomid doesn't work, you can do medicated IUI. If that doesn't work, Plan B might be IVF. And if IVF doesn't work, your next step may be egg donation or sperm donation or adoption. Whatever it is, don't put all your eggs in one basket, so to speak, because the stress really gets bad when your treatment fails and you have nothing to look forward to—that can be utterly devastating.

Before you begin a treatment, talk with your doctor about her long-term plan. Ask her, "How many cycles of this will we do before we move on to the next step? And what is the next step?" It's so helpful to know that for each cycle that *is* a next step. Having a plan that you've already talked to the doctor about can give you something to look forward to.

Make your Plan B as specific as possible, something like this: "Okay, I got my period. I'm going to allow myself to cry and mourn for a few days and then try to get past the tears. After that we'll take a month off and assess what we're going to do next, but I think we're going to try conceiving on our own for two months and then have another treatment cycle. During our break we'll consult with the urologist about having my husband's varicocele repaired." Having a very concrete plan gives you a way to take some control in a situation in which you feel completely out of control.

Sharon knows the value of a plan.

> *I had a medicated IUI recently that failed. I felt very weepy—I was definitely stressed and sad. But I went to a strong place instead of a weak place because I had a plan of action. I turned it around. That day I made the phone call for a second opinion with a new reproductive endocrinologist. I did cry, but I also took control and made it better. In the end I had something—I wasn't pregnant, but I did have an appointment with a new specialist.*

Use cognitive restructuring together with your Plan B to interrupt any negative tape loops that may play in your head when you're faced with an unwanted period or treatment failure, as my patient Bridget did.

> *I'm in the middle of a medicated IUI cycle, and yesterday morning I got the news that I had only one follicle developing. I was so disappointed! My immediate thought was "This month is a wash—this is never going to work." Then I decided to use cognitive restructuring and really focus hard on that thought. I kept telling myself, "It only takes one follicle. That's what most women get pregnant on, one follicle. Why do I think I need to have ten follicles when all my friends who got pregnant naturally only had one good follicle? All it takes is one."*

Many infertile women fall into the habit of telling themselves after a failure that they'll never have a child, they'll never have a child. Using cognitive restructuring, you can give that thought a more positive, constructive spin, changing it to something like this: "I didn't get pregnant this time, but I have another plan all ready. My doctor says this plan may work. I have other viable options."

Of course, sometimes there is no next treatment step—you've run out of money or patience or time, you've reached the end of the treatment line, and you're calling it quits. In this situation failure can be demoralizing—or it may even be a relief. Having an alternative in mind, such as adoption, can help here. I'll talk more about reaching the end of the line later in this chapter.

Make mini-relaxations a major part of your day when you're waiting for the results of a pregnancy test or your period. Do a mini every time you walk into the bathroom. Every time you think you feel a cramp. Every time you begin to worry. When you're doing a treatment cycle, do a mini every time you feel a twinge in your abdomen—even if it's just a gas bubble from that bean salad you ate at lunch—because every twinge will make you flip out with worry. Do minis constantly beginning on day twenty-three. And make sure you're doing longer relaxation exercises every day, even twice a day, because the carryover effect kicks in to keep you calmer throughout the day.

One of the best ways to treat yourself well during injections is what I call "Godiva therapy." Before you start your treatment cycle, buy yourself a pound of really good chocolates. Before your injection, choose which chocolate you'll eat after the injection. It's a reward for having to undergo it—and if your partner does your injections for you, he gets one as well! While recovering from the sting, you can indulge.

The emotional anguish of going through infertility treatment is often magnified by medication. Although the pamphlets that accompany infertility drugs usually report that psychological side effects are uncommon, most of my patients report feeling at least premenstrual when taking them (and their husbands report that it's like PMS squared!). There was actually a study done several years ago that compared psychological symptoms in women undergoing IVF for themselves with those in women undergoing IVF to be an egg donor. The donors reported far fewer psychological side effects from the medications, suggesting that the mood swings and anxiety experienced by so many infertility patients stem from the stress of having to resort to IVF to try to have a baby, rather than from the medications. In any case, if you find yourself feeling "off" while on fertility drugs, this is a really good time to use the mind/body skills described in Chapter 2. Relaxation techniques can relieve the irritability and anxiety, minis can reduce feelings of panic, and cognitive restructuring can help with depressive symptoms. However, please remember that it is okay to feel lousy during this time. The last thing I want to do is to pressure you to feel like your old self.

Last but not least, pamper yourself. Consider yourself under doctor's orders to hang out on the couch watching old movies, reading racy novels, or napping. If you're doing a cycle of medication, you're going to feel tired and bloated and depressed even before you find out whether the treatment succeeded. Make sure you treat yourself well—this is not a time to deprive yourself or to say no to cravings (within reason, of course). Buy yourself perfume, eat ice cream when you feel like it, take yourself out to see a movie everyone's been raving about. Self-nurturance won't take away the pain of a disappointment or failure, but it will distract you a bit by bringing some joy, however small, back into your life.

The key during times of failure is to think hard about what will help you feel better. What has helped you during other times when you felt a huge sense of loss? What has worked for you during past failed cycles or when faced with a death in your family or other sadness? Think back. Does it help to call a friend you haven't seen in a few years and laughing about old times? Going away for the weekend? Creating a piece of artwork? Losing yourself in a mystery novel? You know best what will help you. And if you find that none of those tried-and-true solutions works, talk with infertile friends or members of your support group to see what has helped them. Experiment with coping devices. It's possible that something you think wouldn't help actually will.

When you do a high-tech cycle, keep this cautionary note in mind: If you start to bleed, it does not definitively mean you're not pregnant. A small number of people do have some nonmenstrual bleeding after infertility treatment. Several years ago one of my patients did a medicated IUI and began to bleed two weeks later. She didn't bother going for a pregnancy test, since she was bleeding, and the following weekend she drank a few cocktails. A few weeks later she said she didn't feel well, and I told her to go for a pregnancy test. She was shocked at my advice, but she took the test. It turned out she was pregnant. She delivered a healthy baby, but by having a few drinks she did put the baby and the pregnancy slightly at risk. Don't assume because you're bleeding that you're not pregnant—it's possible you were pregnant with twins and are losing one of the two embryos. Or you might just be having unexplained bleeding. It doesn't happen often, but it does happen. Have a pregnancy test regardless, especially if it's a medicated cycle.

Taking a Break

Sometimes the best thing to do after a treatment failure is to take a break. I highly recommend that if you get to the point that you're feeling burned out, or if you're offered the next step by a doctor and you're not sure whether to pursue it, or if you've done a certain treatment several times and suffered a string of failures, or if you're thinking for the first time of ending infertility treatment altogether, it's a good time for a break. Breaks offer a refreshing escape from the trials of infertility.

Many of my patients choose to take the summer off. They want to spend a few months eating and drinking and having fun, exercising and going on vacation and lying in a hammock, rather than worrying about when to have the next treatment or how to coordinate cycles with vacations or how to keep medications cool in the heat. Taking two months off isn't going to make a difference as far as your age goes—it most likely doesn't matter, fertility-wise, if you're thirty-seven and four months or thirty-seven and six months. If it does, your doctor will tell you.

Margaret found that taking a break made a difference for her.

I've been trying to get pregnant for two years, and it's been very distressing. After my last IUI, I hit rock bottom. It really forced me to reevaluate my life. I came to the point where I realized that I can't let all

> *this failure guide my whole existence. I can't ride this emotional roller coaster every month and let it continue to consume me the way it has. I decided to take a break and feel good physically. I started seeing a chiropractor, a physical therapist, a psychologist, and a masseuse. I had to reduce my stress and regain the power to feel better. It's been a very healing experience. I feel much better prepared to move on to adoption or more treatment, depending on what we decide to do.*

Some time off also revitalized Evangeline.

> *After two years of treatment I'm taking the summer off and starting again in September. Emotionally it's tough, because I have such an unfulfilled desire for a baby. But physically it's been fantastic. I've been going to the gym every day, which I couldn't do for two years. I lost fifteen pounds—when you're taking hormones on and off for two years, you gain weight. Physically I feel great. I've never felt better. I feel like myself for a change. I also renovated the kitchen. It really helped us to have me redirect my energy somewhere else. You become so focused when you're trying to get pregnant. The world around you just keeps going on, but your time seems to stand still.*

When you take a break, you have a chance to laugh and play with your husband and bring some fun and spontaneity back into your sex life. It can make you feel sexy and healthy and strong again. Some people can't take time off, though, because they'd spend the whole time worrying. And that's okay—if you can't do it, don't push yourself to, especially if you're older and feel very tense about time.

Even taking a month off helps. I usually recommend that my patients take the month off after a treatment failure. It gives you time to recover from the treatment and to let your body get back to normal, to pull yourself together emotionally, to be really sure of what you want to do next, to get close to your husband again, to pamper yourself. I think back-to-back cycles are a terrible idea. But not everyone agrees with me on this—a lot of patients and physicians push for back-to-back cycles. You get your period and start another round of medications the next day. I don't think it's wise physically or emotionally. You need time to mourn the fact that this cycle didn't work before you push on to the next one. You may feel desperate to start treatment again, but give yourself a little time off.

My own gut feeling is that your next treatment will have a higher chance of success if you wait a month, although I don't have data to back this up. There have been studies that show, however, that the month after a medicated treatment cycle, spontaneous pregnancy rates are slightly higher. One patient of mine conceived this way. She had two failed IVF cycles—the first was canceled on day twelve due to poor response, and during the second, two eggs fertilized into poor-quality embryos that didn't take after transfer. But she got pregnant spontaneously the next month and delivered a healthy baby.

Surviving Miscarriage

A miscarriage is the cruelest loss of all for infertile couples. You try and try and try to conceive. Finally you succeed. Hooray! You're pregnant! You share your good news with family and friends, you run out and buy *What to Expect When You're Expecting*, you may even purchase maternity clothes. You think you're out of the infertility woods. Suddenly one day you start to bleed. You rush to the doctor, and she delivers the heartbreaking news: You've had a miscarriage.

Miscarriage is a very personal issue to me because I had a miscarriage several years ago. Everything seemed fine at the beginning of my pregnancy. My human chorionic gonadotropin (hCG) level was slightly on the low side of normal, but it had more than doubled every two days, which was reassuring. My doctor basically told me I was fine, and things looked good. About five weeks into my pregnancy I went off to New York to lead an infertility retreat. I went to the ladies' room during a break and couldn't believe what I saw: spotting. I was beyond frantic. Looking back, I don't know how I was able to continue leading the retreat. After that, every time I went to the ladies' room, I was afraid my heart was going to jump out of my chest. And every time I saw blood.

Intellectually I knew that bleeding during pregnancy doesn't have to mean that you're going to miscarry. Bleeding is one of the great myths of miscarriage—you think that blood equals miscarriage. But it doesn't: Women who deliver perfectly healthy babies sometimes bleed during pregnancy, and women who miscarry sometimes see no blood at all. And that's exactly what I would have told one of my patients at that moment. But it was happening to me, and I was scared.

On my way home from New York my bleeding increased. Back in Boston my doctor did an ultrasound, and things looked good—the baby was the right size, and nothing looked abnormal. The doctor reassured me that bleeding can occur in a healthy pregnancy. But I had a feeling. . . . The bleeding continued, and I miscarried two days later. I was devastated.

From my years of counseling couples who had experienced miscarriage, I thought that I knew exactly what miscarriage would feel like, emotionally. But I didn't. I certainly knew intellectually that miscarriage was a risk; my own sister miscarried her first pregnancy, and I was with her in the emergency room. When I miscarried, I was overwhelmed with feelings of loss. What surprised me during the following days and weeks was the sadness we felt. I had known I was pregnant for less than two weeks, yet it took far, far longer than that to start to feel better.

I'll tell you what got me through that terrible time, though: the love and support of my friends and family. For years I'd been telling couples about the importance of social support, but, boy, did I find out for myself how incredibly healing it can be. People came out of the woodwork to tell me that they understood what I was going through because they, too, had had miscarriages. You don't find out until you miscarry how many other women have experienced it. It's amazing. One of my best friends let me know that she'd had a miscarriage and had never told me about it. Everywhere I turned, people were sharing their stories, and that really helped me.

Miscarriage is an extremely personal reminder that the things you love can be lost and that sometimes, you are profoundly unable to protect your loved ones. Most infertility patients live for that first positive pregnancy test. They fixate on day twenty-eight of their cycle—if they can just get pregnant, all their troubles are over. But that's not true, unfortunately. And in fact, 99 percent of my infertility patients find that once they get pregnant, they are shocked at the fact that their anxiety skyrockets. It never occurred to them during treatment that one of the first thoughts in their minds after finding out they're pregnant would be "Oh, my God, I could lose this baby." Miscarriage after infertility is incredibly brutal. It is the loss of an extraordinarily desired baby. And it also represents a loss of innocence for you and your husband. Once you have a miscarriage, you know that you can get pregnant, but you know also that you can lose your baby. It can be given to you, but it can also be taken away. For a woman who has never lost a loved one before, a miscarriage can shake not only her faith that she will become a mother but her sense of overall security in the world.

Miscarriages occur far more often than most people realize. Studies have shown that as many as 75 percent of all embryos that are conceived naturally miscarry. That number seems astonishingly high, but I don't doubt its validity. Researchers using very sensitive tests have found that it's not uncommon for a woman who conceives naturally to miscarry very shortly after conception. In such cases a woman conceives, miscarries a week or two afterward, and has either a normal period or a period just a few days later than expected. She may think nothing of it—for most people a late period is not that unusual. She doesn't even know that she was pregnant or that she miscarried.

Are women who undergo high-tech infertility treatment more likely to miscarry than those who conceive naturally? Researchers don't know the definitive answer to that question, but they believe that the answer is no. It may just seem to be the case because pregnancies are detected so much earlier when women are being treated for infertility. In most instances, for someone with a confirmed pregnancy, even if she has had several miscarriages, the odds are still in her favor that she will go to term. (A confirmed, or clinical, pregnancy means that her blood level of hCG was positive and in the normal range, that if tested again it doubled every two to three days, and an ultrasound confirmed a pregnancy in the uterus.)

It's a good idea to prepare yourself for the possibility of miscarriage.

Not that you should expect it—even if you've miscarried before, you're more likely to carry the baby to term than to miscarry—but taking some preparatory steps can help. For example, don't run out and buy maternity clothes and pregnancy books the minute you test positive, as my patient Cindy did.

> *When I miscarried I had already bought maternity clothes, so I was on all the mailing lists. I got cards for months saying, you're six months along so you must be needing this, or you're eight months along so you must be needing that. My baby was due in September, and in the month of August I got a load of formula coupons in the mail. Then the next month I got all this stuff about newborns. Then you get more mail a year later because you're still on this awful mailing list. It never ceases. It's been two years, and I'm still getting stuff. Now I know if it happens again not to do that.*

Talk with your doctor about how things look, medically. How is your hCG level? In most pregnancies the hCG level on day twenty-eight is a relatively reliable indicator of the health of the pregnancy at that point. If your hCG level is above one hundred on day twenty-eight, it's more likely to be viable than if your hCG is under a hundred. In healthy pregnancies, hCG levels double every two or three days. If yours hasn't doubled, it's an indication—but by no means a guarantee—that you might miscarry.

In most cases you can begin to rest easy after about eight weeks. After a high-tech treatment your doctor is likely to ask you to come in for an ultrasound at six to eight weeks. If at that point your doctor sees a good, strong, normal heartbeat, you have over a 90 percent chance that the baby will survive and you won't miscarry.

Until that point take it one day at a time, especially if you've had a miscarriage before. Don't think about how you're going to survive until you have your ultrasound at eight weeks—think about getting through today.

And think about seeking support from understanding friends and family during this time. A lot of my patients wait until the end of the first trimester to announce their pregnancies. That is the best choice for some people, although these days you can probably feel safe telling people after you've seen a normal heart rate during your eight-week ultrasound. But if you don't tell anyone you're pregnant, and you miscarry, you lose out on the enormous support you could have gotten from friends or family. If you thrive on support from the people you love—as I do—think about telling them. If I had not told anybody that I was pregnant, when I miscarried I would have missed out on all that love.

Here are some other ways to help soften the terrible impact of miscarriage:

- **Give yourself time to grieve.** I strongly advise you not to go right back into another treatment cycle immediately after miscarrying. You have to give yourself and your husband time to mourn. Even if you miscarried very early, it's still a tremendous loss. You created a life together, and now it's gone. Denying the effect that such a loss can have on you—just closing the door and moving immediately on to another treatment cycle—can be detrimental to you in the long run. I can't tell you how many of my patients who tried to sweep the pain of their loss under the rug found that pain resurfacing months or even years later.
- **Seek out support.** I hope that your friends and family will rally around you as mine did. But that doesn't always happen. If not, seek out a friend who can help you through this difficult time. Turn to the women in your infertility support group. My patients often find that the largest comfort comes from other infertile women who have miscarried. They really know what you're going through, because they understand the double pain of miscarrying after infertility, as Eloise discovered.

Miscarriage was the worst thing that ever happened to me. It's just horrible—you're afraid to go to the bathroom because you might see blood. You don't drink water because you don't want to go to the bathroom. You feel like you're going insane. But when I joined a support group, I found that other people had these feelings, too. Before I met these women, I felt like I didn't have a right to feel sad about my miscarriage in front of other people. But seeing that other women felt bad, that it was OK to have these feelings, that it was OK to let other people know you feel bad—that was a big help. I realized that not only did I have a right to feel that way, but that I was normal. It was abnormal for me to be so stoic about it, in fact.

- **Ask for answers.** Depending on how far along you were, your doctor may have some ideas about why you miscarried. Knowing that the baby had major genetic defects, for example, or that the heart never developed won't make the miscarriage any easier, but it may help you not to blame your own body for what occurred.
- **Change the way you relax.** If during your long relaxation sessions you usually choose unstructured methods such as meditation, you might want to consider switching for a while to relaxation techniques that are more structured and give you less opportunity to dwell on your loss. Try yoga, progressive muscle relaxation, mindful walking, and guided imagery. If you listened to a particular relaxation tape while you were pregnant, switching to a new tape for a while may help because you may associate the sound of the old tape with pregnancy.
- **Don't underestimate this loss.** Some people may tell you that you should be able to get over a miscarriage quickly. And perhaps you will. But don't berate yourself if you continue to feel very sad for a while. A miscarriage is a real loss, despite what insensitive friends may say. If you find yourself growing debilitated over the loss of your unborn baby, however, talk with a therapist to see if treatment is necessary.
- **Get ready for thoughtlessness.** Speaking of insensitive friends, you're likely to hear some pretty tactless comments from people regarding your miscarriage. Expect comments

such as "At least now you know you can get pregnant" or "You can always have another" or "There was probably something wrong with it anyhow—it's better this way." When people say hurtful things, you can try to be ready with a response, such as "How is that supposed to make me feel?" or "Yes, this shows I can conceive a baby, but it doesn't prove I can carry one" or "In my mind, this was a baby, and I've lost my baby. How would you feel if you lost a child?" Or you can use it as an opportunity to educate the person by saying something like "That comment really hurts my feelings, and I'll tell you why." Then explain how emotionally wrenching a miscarriage can be.

Renee was amazed by one comment in particular.

A lot of people try to fix your pain by saying things that they think will make you feel better. When I miscarried, a friend of my husband's said, "Well at least you're getting a new car. That should make you feel better." I couldn't believe he thought a new car would make up for losing a baby. I would gladly have driven an old car for another ten years to keep from losing that baby. It doesn't make up for what I've been through.

DeeDee, who miscarried four times, has probably heard them all.

I was just stunned by the things that people said to me after my miscarriages. As if it was an ant or a bug that you could squash and it was no big deal. Like it wasn't really a person. But when you see that baby on the ultrasound, with its heart beating, it's pretty real. People don't understand that. With my last miscarriage, when the baby came out of me, I could see what it was even though it was really tiny. It's a little teeny sac. You can hold it in your hand. Here is your child, and you're putting it in a plastic bag in the refrigerator overnight so you can take it to the doctor's office the next morning. It's so frustrating, because you want to be a parent and you can't protect your child. You're doing everything you can, and they're still losing their lives. It's just so difficult to be unable to do anything to save your child.

- **Honor your baby's memory.** Having some kind of a private ceremony can help you pay tribute to your baby and create a sense of closure to your mourning. When you're ready, think about whether this would help you and what you'd like to do. After my miscarriage, we planted a bush in our baby's honor. After the planting, my husband and I talked about how sorry we were that the baby had died and what our hopes and dreams for the baby had been. That plant is really important to us now, and when we go on vacation, we make sure to ask a neighbor to water it.

  One of my patients had three miscarriages. She and her husband placed candles surrounded by sand into three paper cups and wrote their babies' names on each of the cups. They read some poetry, played some music, placed each of the cups into a stream, and then watched them as they floated away. As the cups drifted downstream and away from them, the couple was able to say good-bye to their children. They were able to let go. Myra and her husband did something very similar.

  My husband and I had our own little ceremony, just the two of us. I bought a candle and a little cross for each of my miscarriages, and I painted a coffee can white. My husband picked out some music, and we played it while we lit the candles and had a little ceremony. Then we put the candles and the cross in the can, along with a piece of paper with the babies' names, the dates they were conceived, the dates they would have been born, and some Bible verses. In the spring, when we move into our new house, I'm going to bury it in the ground.

  The idea to do something like this had passed through my head, but I blew it off because I thought it was a little over the top. But then someone in my support group suggested

> *commemorating my miscarriages in some way. So we did it. It was really good for us—especially for my husband, because he really broke down and admitted for the first time how upset he was about the miscarriages.*

- **Write a letter to your baby.** Tell her how much you love her, how sorry you are that she didn't survive, how much you miss her, what your hopes and dreams for her were. Bury the letter with the baby's remains or with a bush or tree you plant in her honor.
- **Steel yourself for your due date.** Even if you're coping well with your loss, bear in mind that as the miscarried child's due date approaches, your grief may return. If other women you know became pregnant around the same time you did, you may find it hard to be around their babies. Understand that your due date may bring great sadness and that if it does, you're experiencing a very natural occurrence. You might want to mark the date with some kind of ritual. One of my patients who had experienced a stillbirth baked a birthday cake on what would have been her son's first birthday. She and her husband spent the evening together, holding hands, talking about what their son would have been like on that day, and then ate the cake in celebration of their love for him.

Renee was careful to be protective of herself.

> *I had to expose myself very slowly to friends of mine who'd had babies around the time my baby was due, just to make sure I could handle it. I have no problem being around babies in general, but any child that is within about three months of how old my baby would have been—that's tough. It's too much of a reminder of what could have been.*

Think about what will help you feel better on that date—commemorating the baby in some way? Going away for the weekend? To ease stress around that time, increase your relaxation practice and indulge in some extra self-nurturance. By the same token, if you don't feel any particular sadness on your due date, don't feel guilty—everyone reacts differently to such things.

- **Recognize that your husband's reaction may differ from yours.** As with infertility, a woman tends to react very differently to miscarriage than a man will. Because the pregnancy took place in her body, a woman often has a far greater emotional attachment to the fetus than her husband does. When a woman is pregnant, she is aware almost from the very beginning of physical changes—she may be tired or nauseated, or her breasts may be sore—she has physical evidence that she was pregnant because she feels different. For her to miscarry is a very real loss, a tangible change in her body that may be accompanied by pain, bleeding, and other physical events. There was a baby growing inside her, and now there isn't. But for a man it's completely different. Pregnancy doesn't seem quite real until he lays his hands on the mother's belly and feels the baby's kicks during the second trimester. So if your miscarriage occurs before then, your husband probably hasn't connected with the baby as you have, and both the pregnancy itself and the miscarriage are likely to be somewhat abstract to him. This can cause a huge divide between the two of you—you're mourning something real, and he's dealing with something intangible and hypothetical.
- **Write about your loss.** Journaling allows you to explore your feelings of sadness, to mourn in your own way, and to put into words your tremendous feelings of loss. It can also allow you to vent any anger you may feel toward your husband for reacting to the miscarriage so differently from the way you are. This was the case for my patient Susan, who used journaling as a way to let go of the feelings of resentment she felt toward her husband.

A year to the day before my miscarriage, we had to put our dog down, which was tremendously hard. My husband cried and cried and cried—I've never seen him break down like that before. But when I had my miscarriage, he hardly seemed upset at all. I felt like he grieved more for the dog than he did for the miscarriage, and I was really amazed at how resentful I was of that. And I didn't realize until I started journaling exactly how much resentment was there. Now I know that if I have little resentments along the way, I should let them out and I'll be okay. Now I know to tell him these things.

Calling It Quits

Patients are always asking me, "When do I stop treatment?" When finances are an issue, you basically have no choice; you stop when you've spent all that you can or want to spend. But if money is not a factor because of medical insurance or other reasons, it's harder to know when to stop.

When you first start infertility treatment, you have a lot of hope. You're enthusiastic because you assume that your doctor will have something in his bag of tricks to help you conceive. You actively work toward getting pregnant. Sure, the treatment is a pain, but you may not mind too much. It's all for a good cause. After a while, however—and the time frame on this varies from woman to woman—treatment becomes a real chore. You dread the medication. You have nightmares about the side effects. The thought of starting another cycle makes you want to scream. You just can't imagine dragging yourself into the doctor's office for one more blood test or ultrasound or injection.

That's when it's time to think about calling it quits, in my opinion—when you get the I-just-can't-take-it-anymore feeling.

Unfortunately, it's not always that simple. Husbands and wives often disagree on when to stop. Usually the husband is ready to stop before the wife is. He feels they can't afford to continue treatment, and he's reluctant to raid retirement savings, particularly if he's older—it's terrifying for most men to see money hemorrhaged away by infertility treatment. What's more, he's tired of seeing his wife a mess, both physically and psychologically. He may not feel as strongly about having a child as she does, and he wants his life to return to normal, the way it was before infertility intruded. Most of all, he wants his wife back. He misses the healthy, happy woman he married. He may be tired of the depressed, bloated, socially isolated woman she's become.

Sometimes the wife wants to stop and the husband wants to keep plugging along, as is the case with my patient Evangeline.

> *I've had three IUIs and three IVFs. I'm going to do one more IVF in a couple of months, and then I think that will be it. I feel personally that I can't do this anymore. My husband seems to think we're just going to go on and on until it happens, but I just don't know about that. It's hard, because I would love to adopt, but for him it's just out of the question. This is probably going to be my last cycle. That makes me sad.*

Evangeline is in a tough situation—she doesn't want to subject her body to any more treatment, but her husband isn't ready to give up hope. This happens a lot when the cause of infertility is a male factor, and the wife would prefer to move on to sperm donation rather than pump her body full of medication for another cycle. It's a difficult situation because the wife's body takes all the abuse—the only thing the husband has to do is fill a jar in the doctor's office. But the husband is reluctant to give up his one chance of being a genetic father, and he's unenthusiastic about using donor sperm. In other cases the woman is ready to move on to adoption but the man has no interest in adopting.

I see some horrific dissension between couples on this issue. The one who wants to stop feels that the other is totally obsessed with having a baby. The average infertile couple has been going through treatment for three or four years. They may have spent a huge amount of money, endured countless procedures and dashed hopes, and at the end of it all what do they have? Nothing but an empty bank account.

If you and your husband disagree on when to stop treatment, I urge you to see a therapist familiar with infertility issues who can help you resolve your differences and who can suggest ways to make decisions that are palatable to both of you.

Of course, there are situations in which the doctor recommends ending treatment. Infertility researchers are learning more and more about who is a good candidate for treatment, and doctors—ethical ones anyway—will tell you if they think your chances are dwindling. Basically, there are several parameters, such as having a high FSH level or responding poorly to medications over several cycles. Some clinics are stricter about these parameters than others. For example, most programs will not transfer a donor egg into a woman over age forty-six, but clinics in New York and Italy will accept women for donor-egg procedures into their fifties and beyond.

Your doctor may recommend that you end treatment if you're what's known as a "poor responder," meaning that you may have good FSH and estradiol numbers but for whatever reason you don't respond well to IVF medications. Some 20 to 30 percent of women starting an IVF cycle respond poorly to the medications, and they don't have enough follicle or egg development to warrant retrieval. This can be devastating, because if you respond poorly, your treatment will be canceled midway through the cycle, usually by day twelve. It's a horrible shock—people go into IVF cycles thinking they'll have a result on day twenty-eight. To be told on day twelve that the treatment is being canceled can be enormously disappointing. When this happens, the doctor may suggest switching to a medicated IUI, but they tend to be unsuccessful; it's rare to get pregnant on a canceled cycle. Some women who respond poorly to the first IVF do well the second time around, but others do not. In these cases the physician may tell you IVF is not a viable treatment option.

Some couples want to stop treatment because they fear that repeated exposure to infertility drugs may increase the wife's risk of diseases such as ovarian cancer in the future. Is this a valid fear? Well, I can't tell you with 100 percent certainty that infertility drugs are safe over the long term. I can tell you, though, that a very large epidemiological study published in 2002 by Johns Hopkins University found no evidence that infertility drugs raise the risk of ovarian cancer. Researchers will continue to investigate whether there is a connection between infertility drugs and disease, but right now we have no evidence that infertility drugs are unsafe.

Stopping treatment—whether because you and your husband decide you want to stop, because your doctor tells you your chance of success has dwindled, or because you just don't want to expose yourself to any more treatment—can be incredibly difficult. But it can also be a relief to stop fighting with your body and trying to force it to do something it can't do. At this point some couples move on to adoption. Others choose to use donated sperm, eggs, or embryos to create a child who is not genetically connected to both parents. Still others will decide to remain childless.

What will you do if and when you reach this point? In the next chapter I'll talk about all of these options and which may be right for you.

CHAPTER ELEVEN

Other Paths to Parenthood

You and your partner are not the people you were when you first set out on your journey to become parents. Your outlook on life and your perception of what your future will be like have probably changed quite a bit. You imagined conceiving naturally, perhaps having a few children, and now, years later, your dream has not come true. But that doesn't mean you can't become parents. As I tell my infertility patients, if you want to be a parent, somehow, some way you will.

Before you can get to the point where you consider other parenting options, however, you have to be able to let go of the idea that the only child you will accept is one who is formed by the union of your egg and your husband's sperm, carried in your body. This task is more difficult for some people than for others. I have found, however, that with time, a period of grieving, and, sometimes, some good counseling, most infertile couples do eventually decide to move on to adoption or gamete donation.

There are so many different ways to create a family. You can adopt, either domestically or internationally, privately or through an agency. You can use a donor egg, donor sperm, a donor embryo, or a surrogate. Maybe you can't fulfill your dream of delivering a child that is genetically connected to you and your husband, but you can take other pathways to parenthood. And although those paths may not be your first choice, let me tell you something, based on the joy I've seen among my patients who choose these options: These second choices are not by any means second best.

If you don't believe that you can accept adoption or gamete donation—if you're thinking as you read this, "Forget it. If I can't have my own baby I don't want any baby"—I urge you to push those thoughts aside for a short time and read this chapter anyway. Even if the idea of adoption depresses you, if egg donation seems unnatural and surrogacy seems freaky, I still encourage you to read on. Here's why: Very few people start off liking these parenting alternatives. But over time most couples come around to being okay, and eventually very happy, with these choices. They read, they discuss, they think, they let it simmer. They talk with other couples who have taken these paths. They read some more and think some more and agonize some more. Gradually their discomfort turns to acceptance. Their acceptance turns to excitement. And when they hold their baby in their arms—their adopted baby or their sperm-donor baby or their egg-donor baby or their gestational-carried baby—their excitement turns to sheer, unmitigated joy. Finally, after all those years of dreaming, they are parents. Maybe not genetic parents, but real parents nonetheless—the people who will love and raise and *parent* that child forever. You may think of the word "parent" as only a noun, not a verb, but I disagree—in my mind it is the most perfect of verbs. What matters most is not where a child came from or whose genes she's carrying around inside her, but who will love her, who will hold her all night long when she has a 103-degree fever and a double ear infection, who will help her with her math homework, who will walk her down the aisle on her wedding day. Her *real* parents are the people who *parent* her.

But embracing nongenetic parenthood is not a quick or easy process—how could it be? I compare it to falling in love. Most women don't fall head over heels in love with a guy on the first date. You have to get to know him. He may seem nice enough in the beginning, but there are things about him that bother you—the shape of his nose, his annoying best friend, the fact that he doesn't like sushi, his addiction to ESPN. But as you spend time with him, your fondness grows. You begin to accept that he varies from your concept of the ideal man, and you learn to overlook his minor failings and odd personality quirks (the best friend is another story). After a few weeks or months or even years, you gradually fall in love. You decide he's the one for you. No, he's not perfect, but he's yours, and you love him despite his imperfections, *despite his distance from the ideal.* It's the same with nongenetic parenthood. Very few people love the idea right away. I've never, ever heard of a woman who gets her period the first month she's trying to conceive and says, "Excellent! I'm not pregnant! Now I can adopt!" Learning to accept the idea of nongenetic parenthood, despite its shortcomings, despite its distance from the ideal, is a gradual process filled with emotional give-and-take. Just as there are parts of your husband that, despite your love for him, you'd change in a New York minute, there are some things about nongenetic parenthood that are hard to accept. But in the big picture these are of microscopic importance, just like your husband's ESPN problem, because your ability to love is vast and resilient.

I have a pretty good idea of how much you want a child. Just the fact that you picked up this book and have read all the way to the last chapter proves to me that your heart and soul are dedicated to the idea of being a parent. That's why I believe you owe it to yourself to consider all the options out there. I'm not telling you that you have to adopt or use a donor—I'm the first one to admit that those choices are not for everyone, and if you're firmly against the idea of a nongenetic child, *even after lots of careful thought,* you're smart not to pursue it. But have you truly given it lots of open-minded, careful thought? Or have you just pushed the idea aside automatically, without really analyzing why it doesn't appeal to you, or without giving yourself space to warm up to it and, over time, learn to embrace it? If it's not right for you, you can't force it, any more than you can force yourself to fall in love. But maybe it *is* right for you, and you just haven't given yourself the space to realize it, as was the case with my patient Naomi.

I was married at forty. We immediately started trying to conceive, but I had a sense that I might have trouble, what with my age and all. My

FSH level was good, but I opted for medical intervention and began infertility treatment. An IUI failed, so I moved right to IVF, and I conceived the first time around. It was an ectopic pregnancy, though, and I lost the baby and one of my fallopian tubes. I had several more IVFs and conceived twins with the fourth one. But at twenty-one weeks my cervix opened and the twins were born prematurely. They died a couple of days later. It was utterly devastating for me to lose those little girls.

I got pregnant again, with my fifth IVF, but miscarried very early in the pregnancy. Then my sixth IVF failed.

During this time I was completely against the idea of adoption. Adoption felt like it would have been a booby prize—I just couldn't do it, especially when I knew I had the capacity to conceive and carry healthy babies. Adoption would have felt like failure to me. Having a single adopted child would have meant settling for so much less than I had originally had with the twins. Not only would I be getting a nongenetic child, but I'd be getting just one baby, not two. I was willing to keep on doing infertility treatment. It was addictive, because I'd had several successes—I remember thinking I would continue treatment for as long as possible. I just had to succeed again.

A cousin of mine is an adoption attorney, and I had once said to her—just kiddingly—that if she ever came across twins that were placed for adoption, to call me. Well, one day she did call me. She had twins available if I could jump through all the legal hoops fast enough. All of a sudden I became so excited about the idea of adopting twins. I pulled some strings with a friend and arranged an immediate home visit by an adoption social worker, and I worked twenty-four hours a day to try to get that adoption to come through. But it didn't—we just couldn't make the arrangements work.

After that I thought, if I could be so excited about adopting twins, why couldn't I get excited about adopting a single baby? I kept thinking about it, and eventually I came around to the idea. Plus, I told myself that adopting didn't have to mean putting an end to our quest for a genetic child. We could do both if we wanted.

We adopted our daughter, and now I can't imagine any other child being our child. We don't look at her as adopted, we think of her as having come from our bodies. She is so magnificent.

When my daughter was a few months old, we decided to adopt a second child. And then something amazing happened—we received a call telling us that my daughter's biological mother was pregnant again and wanted to give the baby up for adoption. Same mother, same father—we would be able to adopt our daughter's sibling. Now we're waiting for the arrival of our daughter's brother or sister. I feel very strongly that I am my daughter's mother, even though she's adopted. The mother is the person who gets up and burps the baby in the middle of the night, not the one who gives birth. To really earn your Mother's Day card, you have to be there day in and day out. And I am.

In the pages that follow, I'll try to help you along on your thinking process. I'll talk about some of the pros and cons of the various nongenetic options, and I'll provide suggestions on how you can start wrapping your mind around the idea of considering these options. I'll also share some stories from patients of mine—some who have chosen to follow these alternative parenting pathways and some of whom have decided not to. My goal is not to convince you to make one choice or another but to expose you to various ideas so that you can begin to explore them in your own mind.

As you think about nongenetic parenting options, be sure to keep up your daily relaxation exercises. You may want to teach your husband how to do them, too. By taking time each day to empty your mind, slow down your body, and elicit the relaxation response, you'll improve your emotional clarity and eliminate some of the mental clutter that can interfere when you're making an important choice. Try to carve out at least twenty to thirty minutes a day for relaxation during this decision-making process or, if you can manage it, double that time for two relaxation sessions a day. And continue with other relaxation techniques that have helped you. The less stressed and anxious you are, the clearer your thinking will be.

Where to Start

The best place to begin your careful assessment of nongenetic parenthood is to look at the reasons you might be against it. I find that couples with infertility often don't delineate or analyze the reasons that they cross certain choices off their option list. But I believe it's an excellent exercise to think deeply about your own personal obstacles. Once you figure out what they are, you can look at them closely—cognitive restructuring comes in handy here—to decide whether they are truly valid and whether you want to embrace them, restructure them, or discard them.

Here are some of the primary obstacles that my patients have come up against when considering their feelings toward nongenetic parenthood. See which ones, if any, apply to you or your husband.

You're not ready to give up on genetic options. Nongenetic parenthood may seem unthinkable right now because you first need to try more treatments or give yourself more time to conceive naturally. That's fine—many people need to know they've tried everything before they can consider nongenetic options.

The most common question I get about nongenetic parenthood is this: How do you know when it's time to stop thinking about having a genetic baby and to start thinking about other choices? And my answer is that you'll know when it's time. When you first start infertility treatment, you're very optimistic that it's going to succeed. But eventually you get to the point where you think, "I can't stand this anymore." That's a good time to start to consider adoption and other options.

People vary widely in how long they take to get to this point. Some are amenable to stopping treatment early or forgoing medical intervention completely, while others spend years in treatment—some of my patients have done as many as ten IVF cycles before considering adoption or egg donation. After five years of infertility treatment, my patient Harriet wasn't thinking about adoption—but her husband was.

Adoption just wasn't on my radar screen. My doctors couldn't find anything wrong with me, and they kept saying that I might get pregnant. But

> *I was getting older, and my husband is seven years older than I. We had to decide—is our dream to procreate, or to parent? My husband just wanted to parent, and adoption was fine for him. "I don't know that our particular mix of genes is any better than anyone else's. Let's just roll the dice," he said. And I agreed, to a point. But for me the thought of being pregnant and bearing a child and having that unique female experience was something I really yearned for. It's a profoundly difficult decision for a woman to let go of that dream. I always thought that being pregnant would be part of my life experience. Now, years later, I still feel an enormous loss at not having had that experience. And I will always feel that.*

After coming to terms with the idea that she wouldn't conceive a genetic child, Harriet and her husband decided to adopt. They now have two adopted children, and Harriet couldn't be happier.

> *I still feel sad that I couldn't experience a pregnancy, but I'll tell you, if someone could turn the clock back now to before my children were born and they could give me a pill that would absolutely guarantee that I could be pregnant with beautiful, healthy, genetic children, but I wouldn't be able to have my adopted children, would I do it? Of course not! Oh, my God, absolutely, positively not. And every adoptive parent I know would say that. This is the way my family was supposed to be formed. These were the children that were supposed to be in my family and in my life. They are my children.*

You haven't mourned. Before you can move on to nongenetic parenthood, you have to observe some kind of mourning period. As soon as you started dating your husband, you may have begun to dream about what your children together would look like, what kind of people they would be. When you get to the point where you stop trying to conceive a genetic child, you have to mourn the death of that dream. It's really hard, but you have to let go somehow before you can progress.

How do you let go? Well, some couples write letters to their unborn children telling of the grief and anger and sadness they feel about never being able to meet them. Others have a ceremony or do some kind of symbolic act or ritual to mark the end of their "trying" time—for example, they choose to go back on birth control or schedule a vasectomy so that they can end the cycle of hope and disappointment that comes each month.

If you can't let go, if you can't stop mourning, you may not be ready to move on. You should proceed to nongenetic parenthood *only after you've dealt with the pain of not having a genetic child.* Don't try to use adoption or gamete donation as a way to soothe your pain. That's not fair to the child or to you. Become a nongenetic parent because you want to parent that nongenetic child, not because you think an adopted baby may take away the pain you feel over the fact that you can't have your "own" child. It won't work.

It represents failure. Moving on to nongenetic options equals failure for many people. And for some, failure at anything carries a huge amount of emotional baggage, disappointment, and loss of self-esteem. Maybe you or your spouse grew up with a demanding parent who insisted that you succeed at everything. Or perhaps you're a perfectionist of your own making, and nothing less than complete victory in all things is acceptable to you. Either way, I urge you to confront these feelings because they're not healthy for you as an individual, a spouse, or a parent. Why is success so vitally important to you? Are you trying to win a parent's admiration? Are you competing with a sibling in a decades-old race to be the favored child? Do you have a compelling need to be more successful than your friends and neighbors? Are you trying, as an adult, to make up for failures of your youth? If your answer to any of these questions is yes, I encourage you to think deeply about your ideas on success and failure, both in general and in relation to your desire to become a parent, and perhaps to seek counseling.

Use cognitive restructuring to put your thoughts of failure to the test. Does your insistence on total success contribute to your stress? Where did you learn this habit? Is it logical to expect perfect performance every time you set out to achieve a goal? Is it true that you won't be happy unless you succeed in everything?

Another way to look at failure is to separate yourself from your body. Perhaps your body is indeed failing to create a child—maybe your infertility is clearly the fault of your ovaries, or your scarred fallopian tubes, or your soaring FSH level. But that's not who you are. You are not your ovaries or your fallopian tubes or your hormones. You are a whole person made not only of cells and organs but of personality and love and spirit. Yes, perhaps a part of your body is failing, but that doesn't make *you* a failure. Try to separate yourself from the cause of your infertility and to accept it as a condition over which you have no control, rather than a personal shortfall.

You think a nongenetic baby will be an imperfect substitute for a genetic child. You desperately want a baby who will inherit the best qualities of both your family and your husband's. Only a genetic child would have your musical abilities, your husband's sense of humor, your father's athletic build, your mother-in-law's curls. This is flawed thinking, because genetics can be a very inexact science, and you have no guarantee that your genetic child will inherit the right traits. Yes, she may have your musical abilities, but she may also end up with your Uncle Pete's crooked nose and your sister-in-law's thunder thighs.

I saw a couple, Pam and Matt, for counseling because they were stuck. She was forty-two, they had tried four IVF cycles—all unsuccessful—and Matt wanted to move on to egg donation or adoption. However, Pam resisted giving up on treatment. When I asked her to describe her desired baby, she said her baby would look exactly like her. In fact, she *needed* her baby to look exactly like her. If you could have seen Pam and Matt, you would realize how completely unrealistic this was. Pam was tiny, blond, and blue-eyed, while Matt was the exact opposite: tall, athletically built, and from a Mediterranean background, with black, curly hair and dark brown eyes. Since dark eyes and hair are genetically dominant, it was pretty unlikely that the two of them would produce a blond pixie like Pam. As it turned out, Pam was extremely close to her mother and sisters, all of whom looked very much alike, and she was convinced that in order to feel close to her own child, that child would have to look like the rest of her family. We spent a number of sessions doing cognitive restructuring, challenging Pam's beliefs that the only child she could love would be one that was her clone, especially given the coloring of the man she loved. Pam realized that the closeness she felt with her family was more about the shared experiences, adventures, and mutual respect than it was about eye and hair color.

Letting go of the expectation that your child will look or be just like you can be painful. Tricia's husband, Jack, was sterile due to chemotherapy treatment for childhood leukemia. They decided to try to conceive with donor sperm. Tricia came to see me for help in choosing an appropriate donor and showed me her top ten picks. To my astonishment, all of the donors she had chosen had her coloring (light brown hair, hazel eyes) rather than Jack's—he's from a Middle Eastern background and his coloring is much darker. I explained to Tricia that if she chose a donor with her coloring, the child would look nothing like Jack. But she was adamant that the child look like her and refused to come back to see me for any more sessions. I don't know what they eventually did, but I'm concerned about Jack's feelings. She loved him enough to marry him but didn't want their child to look like a blend of the two of them.

As you make choices and decisions in this area, think carefully about the expectations you have of your future children and make sure that you don't project your own issues onto them. Many adoptive parents have told me that one of the joys of adoption is the thrill of discovering their child's talents—which are often unique in their family. It is also easier for a child not to feel pressure to, say, be good in math because that is your specialty.

You'll be happy only if you have a genetic child. That's a lot of responsibility to put on a seven-pound infant. Is it really fair to hang your own ability to be happy on the tiny shoulders of a baby? Think about it: Why is it that passing on your genes is the only thing that will make you happy? What fantasies do you have about this child that may not be based on reality? That he'll be perfect because he's made up of your genetic material? That he'll be able to accomplish everything you couldn't accomplish? Those are a lot of expectations for someone so little.

You'll be happy only if you can have an infant. In some adoption circumstances, particularly overseas adoption, the child you receive may not be an infant but a child of a year or more. That's unacceptable to some couples, who can't let go of their dream of having a newborn child. When my patients can't let go of this idea, I remind them that becoming a parent is about having a child, not just a baby. They may fixate on the idea of feeding, holding, cuddling, and changing a tiny infant, but I remind them that infancy lasts a short time and that parenthood is a lifelong relationship that goes well beyond the infant/toddler stages. Yes, you do miss out on infancy when you adopt a year-old child, but think of all of the many joys you'll share for the rest of the child's life!

You don't know if you could love a nongenetic child. What if you adopt a baby and don't instantly fall in love with it? What then? Well, you're right, you may not immediately adore your nongenetic child. But let me tell you something: People don't always fall in love with a genetic baby right away either. It's such a myth that as soon as the doctor delivers your baby you're totally in love with it. That ain't necessarily so. For a lot of people it takes a while to fall truly in love with a baby, genetic or not. It did for Bill.

I remember when we adopted our daughter, the woman who ran the orphanage handed her to us. She was wearing a dress that was way too big for her, and she had some kind of skin irritation that made her skin all pimply. I took a look at this little thing and thought, "This is it? This pimply little thing is our child?" It's funny to think of that time so long ago. My daughter is six now, and I'm so involved with her—back then she was like a completely different person.

Sometimes with adoption and egg or sperm donation, you have to take a leap of faith that you're going to love your baby. It might feel as though you can't trust the process, but I've never seen any parents not fall completely, foolishly in love with their adopted baby—eventually. You have to trust that you will.

One of my patients was quite hesitant to adopt because he didn't know if he could love an adopted child, but he took that leap of faith and agreed to adopt despite his concern. When he and his wife went to the hospital to pick up the baby, he was still worried: What if I don't bond with this baby? What if I don't love this baby? What will I do? They walked into the nursery and looked at all the little bassinets, and wondered which was their baby. When they found their bassinet the husband picked up the baby and looked at her and suddenly realized that he would indeed be able to love her. As he held her, he told her, "Although I will probably never know what it feels like to love a biological child, I know I could never love another child more than I will love you."

You're concerned about what problems a nongenetic child may have. This is a legitimate worry. When you adopt a child or use a donor egg, sperm, or embryo, you may be at greater risk of parenting a child with emotional or medical problems. There are several ways to think about this:

- First, clinics that collect donor egg and sperm screen their donors—at least, the good ones do. You can reduce worries by choosing a clinic with an excellent reputation and a thorough screening process.
- Second, there are ways to lessen your risk of adopting a child with problems. For example, you can choose open adoption, an arrangement in which you can spend time with the birth mother, learn about her family history, and be involved in her obstetrical care.
- Third, no matter what you do, you can't eliminate all risk of getting a child with health or behavioral problems, and if you can't live with that, maybe you should reconsider whether you even want to be a parent. No matter what kind of care a pregnant woman receives or how sterling her health history, she can still deliver a child with health problems. But the same would be true if you gave birth to a genetic child—there are lots of people who have spontaneous, easy pregnancies that do not produce perfect children. You may even have friends or family members who have experienced this. There are no guarantees either way. Accepting risk is part of being a parent.

You're concerned that the birth parents will come back and try to take your adopted child away. States do give birth parents some period of time to change their minds—ranging from a few days to a month or more. You can minimize your risk by working with an agency that provides excellent counseling for the birth mother and that avoids drawing from states with long revocation periods. People who cannot tolerate such risk often explore international adoption. However, it's important to remember that even in some international adoptions the birth parents are involved.

Nobody else you know has done this. One of the most powerful things we do in our mind/body infertility program is to invite a couple that has gone through this journey to come in and talk about their experience. It is mesmerizing to listen to the story of a couple that initially was cool to the idea of adoption or egg donation or whatever else and eventually came around to embracing it. These parents are absolutely crazy-nutty for their babies, and it's wonderful to hear them tell their stories (and watch them fawn over their children). If you don't know anyone who has been down this road, I urge you to seek people out. Ask your friends and neighbors if they know anyone you can talk to, check with Resolve, ask your doctor or clinic social worker or your clergyperson. Nothing beats sharing with those who were once in the same place you are now. They'll answer your questions, reveal their ups and downs, and help you weigh the validity of your fears and worries.

You, your spouse, or someone close to you was adopted. A number of my patients have a close association with adoption. For some of them that makes adoption harder. One adopted patient wouldn't even consider nongenetic parenthood because it was incredibly important to her—more important than anything else, in fact—to have a genetic tie to her child, since she had no genetic ties to her family. Yet having experience with adoption makes some people very open to it.

Your family is against it. Your husband is Harold E. Smith III, and his family would be mortified if a nongenetic child inherited the noble title of Harold E. Smith IV. Or maybe your family members say they won't accept a child from another race or culture. Families can put a lot of pressure on an infertile couple to keep trying to have a genetic child even when the couple is ready to move on. What are your options? Tell your family to mind their own business, and go ahead and do what you want. Or use a sperm or egg donor who resembles you or your husband, and don't tell your family—your child can decide whom he wants to tell when he's older. As you make your decision, think hard about what kind of family you have. I've seen many situations in which members of the extended family claim they will have nothing to do with an adopted child, and once they get that child into their arms they melt like ice cream on a summer sidewalk. Unfortunately, though, I've also seen families reject nongenetic children. Only you know what your family is capable of and how strong you would be in support of your child, faced with stubborn familial disapproval.

There is something else at work here. Some people aren't sure why they're reluctant to adopt or use a donor. In these cases a therapist may be able to help unravel an answer. A while back a patient's husband was terribly resistant to adoption, but he didn't know why. After a string of visits to a therapist, he realized that he hated his career and was uncomfortable bringing a child into his family when he was so unhappy. After he changed jobs, he was much more satisfied with his life, and the couple adopted happily.

You and your spouse disagree. One of the hardest hurdles to overcome is when you and your spouse disagree on nongenetic parenting. If you don't agree, it can be a tough place to be, particularly after enduring years of infertility, as it was for my patient Judy.

> *I'm ready to move on and adopt, but my husband is still trying to get his arms around it. For three and a half years I've been doing nothing but infertility drugs, and I've had enough. We're working through and communicating about how we can get to the next stage and what we can do to be happy. I'm already there, but now I need to bring him along. But I don't want to force him, if he's not ready. I bring it up a lot and talk about it. The way our relationship is, and the way our jobs are, it's easier for me to kind of just blow ahead on this and let him follow when he's ready. I've started doing the legwork—we have an appointment for*

an information session in a few weeks—and he's open to going to these sessions. But we're definitely in a different place. I would be happy either way. If I got pregnant, wonderful, or if I adopted a child, I know I would love that child as much as I'd love my own. I know I'd be happy either way. But my husband has to decide on his own what he wants to do. I can't pressure him, but I can give him all the information he needs in order to decide.

Forcing your spouse to adopt or use a donor isn't fair to your spouse or the child—or to you. A baby has the right to come into a home where both parents urgently want him. If you and your spouse can't agree, I urge you both to follow the steps recommended in this chapter and to analyze deeply why you each feel the way you do. Then, after thinking carefully, discussing the issue, and exploring all options, if you still find yourself at odds, I recommend a few sessions with a therapist, preferably one who has experience with infertility and adoption. This is not an easy problem to solve, as my patient Cassey knows.

My husband just doesn't seem to have the same need for a child that I have. If he's going to have children, he wants them to be his own. I would love to adopt, but for him it's just out of the question. I just want a baby, but he wants someone who is a biological child. I don't think I can change his mind, which is sad.

When my patients in the Boston area find themselves in a deadlock situation with their husbands about whether to move on to nongenetic parenthood, I refer them to Ellen Sarasohn Glazer, an infertility and adoption counselor in Newton, Massachusetts, and author of several books, including *Choosing Assisted Reproduction: Social, Emotional & Ethical Considerations* (Perspectives Press, 1998) and *Experiencing Infertility: Stories to Inform and Inspire* (Jossey-Bass, 1998). Ellen meets with couples and helps them find solutions to their disagreement over whether to adopt or choose another form of nongenetic parenthood. I asked Ellen to answer some questions about how she helps couples, and I thank her for her insight:

Q: What do you tell a couple that comes to you for counseling because one is ready to adopt and one isn't?
A: It's more of an art than a science. A lot of what I do with couples is basically to help them distinguish whether no means not yet or never. When one partner says no and that really means never, that's a problem. But I find that what no usually means is not yet. That person is just not ready.

Q: If no means not yet, what's the next step?
A: I urge the partner who is ready to move on to adoption to have patience. I try to help that person have a lot of respect for the partner who is struggling. So often the wife is ready to move on and the husband is stuck. In my experience the men eventually come around, but they come around only after they have grappled with some very serious issues. They may not want to adopt for a variety of reasons—they have children from a previous marriage, or they are holding on to the dream of having a genetic child. Before he can say yes, a husband has to accept the fact that he could agree to go along with adoption and then his wife could be hit by a truck, and he'd be left as the single parent of a child he wasn't eager to adopt. These are serious issues, and the eager spouse has to have a lot of respect for the fact that her partner is not taking these issues lightly. Usually the man is in a lot of agony. He says, "I just can't be forced to do this. I can't compromise. I can't say yes before I'm ready. This is forever."

Of course, with some couples one partner is just avoiding the issue of nongenetic parenthood and isn't seriously considering all the issues. In a situation such as that, the eager partner may have to push the reluctant partner to stop avoiding the issue.

Q: When a partner who has said not yet finally says yes, is that partner a good parent?
A: Yes. When those partners make the decision, I find it is sincere and wholehearted and quite authentic and durable.

Q: What happens when no means never?
A: That's definitely a problem. Then you have to decide. Do you stay

together? Do you split up? And if you stay together, the eager partner will probably have a lot of anger. She'll probably need counseling in order to defuse that anger. And I would also suspect that in a marriage where one person is saying never and the other is really hurt by that, there are probably other problems that need to be addressed as well. When a man says never—and it is usually the man who says this—there is a tremendous potential for damage to the relationship. Although he does not want to be hurtful, he is denying the woman he loves the ability to be a mother, and that's enormous.

Occasionally my patients find very creative ways to settle these issues. One patient left her husband when he said never. She lived on her own and adopted a child on her own—but she and her husband continued to "date" on weekends. He just didn't want the day-to-day responsibility of raising a child. This is an unusual arrangement, but, hey, it works for them. That's what matters.

Q: What reasons do your patients most often cite for saying no to nongenetic parenthood?

A: It varies. The most common reason is that they're afraid of living with a stranger—they fear that when the biological glue that binds family members together is missing, the relationship will be weak. People also fear getting a defective child, or they worry that something will go wrong—especially that the birth parents will come back to reclaim the child after the adoptive parents have grown attached to him—it's a huge fear, but one that is largely unfounded. Birth parents, for the most part, make carefully thought-out decisions. And state laws protect adoptive parents. They worry about the child's background, since adoptive children may come from high-risk backgrounds. But there are very individual reasons, too. One of my patients had an extremely loving relationship with his father. He wanted to have that same deep relationship with his child, but he was concerned that he wouldn't feel such a strong bond with an adoptive child. At the same time he was frightened that if he said no, he would lose his wife. It was a lose/lose situation for him. Over time he realized that he would rather risk adoption than losing his wife, and they adopted.

Q: But does a partner in that situation make a good parent?
A: In my experience the answer is almost always yes. One of my patients didn't want to be a parent, but he said yes because he saw how much his wife wanted to be a mother. He said, "I love her enough to say yes even though I want to say no." Five years later not only is he the father of two Russian children, but he's the head of a Russian adoptive parents' association.

Q: Some couples can embrace the idea of nongenetic parenthood so much more easily than others. Why is that?
A: There are many reasons, but a primary factor is whether their decision coincides with an end to their dream of having a biological child. If you're thinking about adoption because a doctor has told you that you have absolutely no chance of conceiving, period, you're not only grappling with the idea of having a nongenetic child but you're also coping with the loss and grief of forever closing the door to your fertility. But if you're adopting because you've been trying to get pregnant for a while, haven't succeeded, but may pursue treatment in the future or have reason to believe that you may conceive spontaneously at some point, that's a very different story. Choosing adoption is often more painful for a woman who is forty-three and has been told that her eggs are, for all practical purposes, useless, than for a woman who is thirty-two and has unexplained infertility. The emotional process is very different when there's a potential, however small, for a biological child in the future. It has to do with finality and the nature of the grief.

The length of your infertility experience also plays a part. I sometimes see couples who've been infertile for years and years. They've tried every treatment, they've had IVF after IVF, and still they're not pregnant. I call these people the walking wounded—they're exhausted and depleted and have so little stamina left to face the issues of adoption and to jump through the hoops necessary for adoption.

Q: How can an eager partner help a reluctant partner?
A: That's a tough one. The most important thing to do is to figure out if

no means not yet or never. If it means not yet, then the eager partner just has to step back and be patient while the reluctant partner grapples with his issues. Couples move at very different speeds on important decisions like this, and that's perfectly healthy and acceptable. The eager partner can't push the reluctant partner or make him feel guilty for his hesitation, because not only does pushing rarely help but it can create other problems in the marriage. I highly recommend counseling for the couple or the reluctant partner. The issues involved are very serious, and they won't just go away on their own.

Q: How do you find a good counselor?
A: That's another difficult question. A lot of counselors know nothing about adoption and infertility and nongenetic parenthood, and in my opinion it's crucial that couples work with a counselor who has experience in this area. Check with your local chapter of Resolve, ask local adoption agencies and infertility clinics if they have a list of recommended therapists, and talk with people you know who are infertile or who have adopted children.

Now let's talk about the various nongenetic birth options, because educating yourself on all the alternatives can help open your mind, too. Here are some of the pros and cons of the various choices, based on the experiences and concerns that my patients have had over the years. This is by no means an exhaustive discussion—entire books are devoted to each of these topics, and I recommend that you read widely as you make your decisions. For some suggestions on books, Web sites, and other sources that can supplement this information, see the Appendix at the back of this book.

Adoption

There are several different types of adoption, and each has its pros and cons. Choosing which type of adoption to pursue is an enormously complicated decision, and I urge you to become fully educated on the topic before making a choice. Also, you'll need to think about your priorities regarding the child's race, culture, age, health, and gender, as well as the degree of openness you'd like with the birth parents as you plan your adoption strategies.

I'm not going to go into detail about the different types of adoption, because there are lots of books out there that cover the topic more thoroughly than I could. What I see my patients wrestle with sometimes, however, is the choice between domestic and international adoption. Again, there are pros and cons. Adopting a baby from overseas can take less time than a domestic adoption, but there are issues to consider: You may not know anything about the health of the baby's parents, the mother may not have had good prenatal care or immunizations, and the baby may not have had intellectual or emotional stimulation in early infancy. (Although I have seen many international adoptions result in perfectly healthy children with no more health problems than domestically adopted children.)

Also, you may not feel equipped to raise a child of another race or culture. One of my patients, a white man, couldn't consider anything but a white child—he said it was such a leap for him even to accept adoption, and he couldn't leap any further and adopt a child who looked different. That attitude won't win any awards in the world of international relations, but I credit my patient with knowing his limits. Some people love the idea of bringing another culture into their family, though, and they're happy to take part in reunions that bring together adopted children from a certain country. Agencies that arrange adoptions of Chinese girls often hold social events for the girls and their families that give the adoptees a chance to socialize and connect with others who share their experiences. Some couples love that kind of thing, but others don't. Again, I think it's good for you to know your limitations, but in my opinion, if you're going to adopt internationally, or if you adopt within the United States and choose a child of a different race, you owe it to your child to raise him with some connection to his birth culture. Bill, whom you met earlier, has faced this obligation.

> *We didn't want to get into the competition to adopt a white American child. We didn't want to wait a long time or spend a lot of money, and we asked ourselves if we were comfortable with having an interracial*

family. And we were, in terms of a Chinese child. The most important thing for us was to have a family and to have a healthy child, and from what we learned from friends who had adopted, there were plenty of healthy girls available to adopt in China. It was a pretty straightforward decision, but we had to ask ourselves a lot of questions. Would we be comfortable adopting from China? What would it entail? We had to accept the responsibility we have to our daughter to help her deal with not only being adopted but being adopted from another country. We know that there will be different identity issues that we'll need to help her with over the years. And we're committed to that process. I'm very happy with my family. I can't even conceive—excuse the pun—of not having her. I don't even think about whether she's different from me. She's my child, that's all.

One reason that couples choose adoption is parity. With sperm or egg donation, one parent is genetically related to the child and the other isn't. But with adoption (and with donor embryo, which some couples choose for the same reasons), neither is related. That is soothing to some couples.

Donation

The main advantage to donation is that it allows a woman to experience pregnancy and breast-feeding. Couples can choose to use donor sperm, donor egg, both, or a donor embryo. There are advantages and disadvantages to each of these. The overwhelming advantage of using a donor, as opposed to adoption, is that you have control over so much more than you do when you're adopting. You can control the pregnancy as much as possible by getting excellent prenatal care, taking vitamins, eating well, and refraining from destructive behavior. When you adopt, you don't have control over the choices that the birth mother makes while she's pregnant.

Sperm donation is less common these days than it used to be (except among single women or lesbians) because intracytoplasmic sperm insemination (ICSI), a procedure in which a single sperm is injected directly into a single egg, is so successful. With ICSI, men with low sperm count or poor motility can usually become genetic fathers. It's expensive, though, because it requires a full IVF cycle, which can be costly. For couples with limited financial resources, using donor sperm is sometimes the only option, because it's less expensive than IVF with ICSI. Egg and embryo donation are more common, particularly for women with high FSH levels, as was the case with my patient Celine, who gave birth to twins using donor eggs and her husband's sperm.

> *My doctors decided I wasn't getting pregnant because of an egg issue. They suggested I use donor eggs. There was a long waiting list for unknown donor eggs, and I was forty-two and kind of in a rush. So my brother's daughter donated the eggs. She's very grounded—married, two kids, very fit and healthy and psychologically strong. She and her husband met with the therapist at my infertility clinic, and everything seemed OK.*
>
> *Our whole family thought it was the greatest thing. And my niece was thrilled to do it—she told me that she believes one of the reasons she was born was to do this for us. She is the most amazing woman. My niece is a neonatal intensive care unit nurse, and she came for two weeks after the birth to help out. When she arrived, people said, "Isn't it going to be weird having the mother help take care of them?" But she's not the mother, she's the donor of genetic material. She's the biological parent, not the mother. I asked her if she'd feel strange taking care of them and she said, "I'm certainly more interested in seeing what they look like than I would otherwise be, but that's it. I feel sort of like a grandmother feels—I have a special connection to them, but they're not mine."*
>
> *I admit there's a weirdness factor when you first think about using donor eggs from a family member. What do you tell the children? How do you explain that your aunt isn't really your aunt and your cousin isn't really your cousin? You do have to process all that. But once you do—and we had help from the counselors at the infertility clinic on that—it's OK.*
>
> *As we went through this whole infertility process, every time one door would close, it seemed that another would open. Ahead of time we*

weren't able to look at every door and say, "OK, let's go through them all." But as each one presented itself, we opened it. And once the donor-egg door opened, it seemed like the right thing to do, so we went full steam ahead.

Even though I'm not the genetic parent, the kids have a lot of my genetic material. My brother and I are a lot alike, and his daughter is just like him. I think of these children as mine, through and through. Even though I'm not the genetic parent, I'm definitely the mother.

There are several advantages to using donor sperm, eggs, or embryos rather than adopting. If your insurance covers infertility treatment but not adoption, using a donor will cost you less out of pocket than adoption will. You'll have the experience of being pregnant, delivering a baby, and nursing. And you have the option of keeping your child's genetic source a secret from the world (although I don't recommend keeping it a secret from the child—I'll talk more about that a little later).

But there are disadvantages, too. If one parent is genetically related to the child and the other isn't, the infertile person may feel less of a connection and will worry that he or she won't be an equal parent. This is most pronounced with sperm donation, because the husband is left out, not only of the gestation of the baby, which is normal, but of the conception as well. It's less of an issue with women using a donor egg, because the woman forms a bond with the baby during pregnancy and nursing.

Sometimes a man may feel weird about the idea of his wife's being impregnated with another man's sperm. He may worry that his wife will have sexual fantasies about the donor, the mysterious stranger who "knocked her up"—although usually some honest conversations and reassurances take care of this issue.

Another disadvantage of using a donor is that there's no guarantee that an infertile couple will conceive using gamete or embryo donation, particularly if the cause of their infertility is unknown. This can be frustrating after years of infertility and treatment. Also, there's always the risk of miscarriage, although the risk is no higher with a donor baby than with a genetic baby.

As you think about using donors, here are some of the questions that might arise:

Should we use a known or unknown donor? Using an unknown donor allows you to pick and choose among a variety of donor characteristics, and you won't have the complications of having the donor be a part of your family's life. But some couples want to use a donor who *is* part of their family's life. Using a known donor allows you to be aware of the background of the donor and perhaps even to be genetically connected to your baby, if the donor is a family member.

How should we go about deciding which family member to use as a donor? This is a tough one. The best situation tends to be when a sibling who has already had children volunteers eggs or sperm. Things get stickier when the family member hasn't had children yet—what if she's unable to have her own children years later? You want to choose a person who can handle the emotional aspect of donating a gamete and who won't interfere with how you're raising the child. In most infertility clinics the donors are screened for psychological stability by an infertility counselor.

Lots of other issues arise. What if you ask a family member to be a donor and he says no—will your relationship be damaged forever? What if a family member volunteers but you don't want her to be your child's genetic parent? Would you accept your mother's egg (if she's young enough) or your husband's father's sperm? (These are very touchy options.) What if the ultrasound or amniocentisis shows an abnormality—would the donor fight you if you chose to abort the fetus?

How do we ask someone to be our donor? Mention it in conversation in a general way. "I was reading an article about egg donors last week," you might say to your sister. "It sounds like an interesting idea—what do you think?" Maybe your sister will volunteer then and there. If she doesn't, write her a letter. Enclose a brochure from your infertility clinic about what's involved in being a donor—a man can make a donation in minutes, but for a woman an entire IVF cycle is involved. Explain how the medical care will be paid for—will you cover everything, or will your sister's insurance foot the bill? Will you reimburse her for lost work time? Don't assume she knows anything about egg donation, because many people in the fertile world are unfamiliar with it. Be sure to explain in the letter that it's OK if she says no. Being a donor—particularly an egg donor—is an enormous commitment, both physically and psychologically. Some people wouldn't be able to handle it, and you need to give them the space to say no without guilt.

Do we tell others that we're using a donor? This is a very controversial question, and there are debates at almost every annual fertility meeting on the pros and cons of disclosure. I personally vote no on this one. Most people who use donors don't tell, and I encourage that among my patients. In a study done by University of Connecticut researchers, 75 percent of couples who used donors and told relatives about it regretted telling. When you tell, you risk having people think of you as less of a mother or your husband as less of a father. It can be very uncomfortable. I advise all my patients, "Don't tell anyone right now, while you're considering the option. You can always tell when you're pregnant, or when the baby is born, or when the child is in kindergarten. But you can never untell."

That said, however, I firmly believe that the child should know the truth about his genetic origins. Research shows that when kids are not told, they often find out anyway, and if they find out when they're older, it can be devastating to them.

Adoption studies support this—when parents are matter-of-fact and up-front about adoption, kids do much better. I believe the child should be told as soon as he's old enough to understand, and then he should decide whom to tell. This goes for children who are conceived with the use of assisted-reproductive technologies. I feel very strongly that children are entitled to know where they come from. An excellent book that addresses this topic for children is *Mommy, Did I Grow in Your Tummy? Where Some Babies Come From*, by Elaine R. Gordon.

When do we begin the donor process? In some states there are long waiting lists—eighteen months is not uncommon in states that have insurance coverage—for donor eggs and embryos (but usually not for donor sperm). If you think that using a donor is one of the options on your Plan B list, put yourself on the waiting list early. Then, when the clinic calls you to let you know a donor is available, be very honest with yourself—are you really ready to do this? And if you're not, take a pass and ask to be put back at the end of the list.

Two things to keep in mind: First, if you put yourself on a donor list, you may have to do so at another clinic, because most won't allow you to undergo treatment and occupy a slot on the donor list. Second, be aware that some clinics ask for a deposit when they put you on their donor list, and if you decide not to use the donor egg or embryo, you lose your deposit.

Surrogacy

There are two kinds of surrogacy: In traditional surrogacy the surrogate becomes pregnant with her egg and your husband's sperm. This is risky, because the courts are divided on what to do if the surrogate changes her mind—flash back to the case of Baby M, who in the mid-1980s became the subject of a court battle between her genetic mother, surrogate Mary Beth Whitehead, and her genetic father and his wife. The second kind of surrogacy, which is much more common, is gestational surrogacy. An egg (either yours or a donor egg) and sperm (either your husband's or donor sperm) are joined via IVF and transferred to the uterus of a surrogate. This is called gestational surrogacy because the gestational carrier, although she carries and delivers the baby, is not genetically related to the baby.

Surrogacy is very expensive, costing in the neighborhood of thirty thousand to sixty thousand dollars for the IVF cycle, the attorneys' fees, the surrogate fees, and health expenses, plus a bonus for twins or triplets. That puts it out of reach for most patients, unless a friend or family member volunteers her services. In that case the situation is a little less dicey, because your surrogate isn't genetically connected to the child. But she will be making an enormous commitment to carry a child or children for nine months and then deliver either naturally or via C-section.

A lot of what I said about known donors applies to family members or friends as gestational carriers—how you approach the topic with them and so on. What changes is the magnitude of the sacrifice a surrogate makes for you. If you use an unpaid surrogate, be sure that she is someone whose heart is big enough to make such a sacrifice—and that yours is big enough to appreciate it.

No matter what form of nongenetic parenthood you choose, if it is really what you want and feel comfortable with, that's the path to parenthood you need to take. I receive many calls from patients who have adopted or used donors, and they tell me that they're so happy with the way things turned out. Many of them go so far as to tell me that they're glad they never got pregnant naturally, because if they had, they wouldn't have ended up with the child that was clearly meant to be theirs. Even people who are not religious say that fate or God or somebody brought them and their baby together, and that was the way it was meant to be. My patient Millie is one such person.

I love my adopted sons more than life itself. I'm glad I didn't get pregnant because if I did, I wouldn't have these boys. Don't get me wrong—I still feel terribly sad that I missed being pregnant, feeling labor, and nursing. In fact, six years later I still feel a twinge of envy when I see a pregnant woman. But I wouldn't trade my adopted boys for anything, including a genetic child conceived naturally.

Choosing Childlessness

This is the path Christine chose:

We tried for six months to get pregnant naturally, but nothing happened. I tried Clomid, I had three IUIs, and then it was time to decide whether to move on to IVF. I thought about it long and hard, and I simply couldn't see shooting myself up with all those hormones repeatedly. It just didn't feel right. My husband would have gone along with it, but I decided not to do it. So we're going to remain childless. We did think about adopting. I talked to people who were adopting children, and I knew they were comfortable with it. But I wasn't, for a number of reasons.

A woman in our mind/body program announced during class one day that she was pregnant, and I remember being really happy for her. One of the other women said, "I can't believe you can be so happy!"

and I said, "I think she needs a baby more than I do." Some women just need a baby so much. Their lives are crumbling without one, and their marriages are falling apart. I didn't want that to happen to me. Maybe because I got married later in life, I put a really high value on my relationship with my husband. I wasn't willing to jeopardize that for a baby—I wasn't willing to ruin my whole life for it. I think a lot of women put all their happiness in danger for this ultimate prize, this ultimate gift. I didn't want to do that.

I think I've made peace with my decision, although I don't know that I'll ever make total peace with it. It's something I wanted very badly, and I'll always miss it. But as badly as I wanted a baby, I never thought of it as something I had to have, something I wouldn't be able to move forward without.

I took the point of view that not everyone gets everything in life. I've been very lucky that I have a great family and a great husband and great friends and a really good life. This is the one component that would have been the total icing on the cake. I guess that was how I was able to make peace with it—that we all don't get what we want in life, and sometimes you just have to accept it and move on. That's what I'm trying to do.

After lots of thinking and analyzing and talking and reading, you may decide that nongenetic parenthood isn't the right choice for you. In that case your options are to continue infertility treatment or to choose to remain childless. This is a terribly difficult decision, but one that sometimes feels right to infertile couples. They try it on for a while, usually during a break from treatment, and find that it feels comfortable. They may declare an end to trying to conceive, or they may just stop treatment, let nature take its course, and keep in the back of their mind that if childlessness doesn't feel right, then down the road, perhaps they can adopt.

Bridget finds herself currently in that position.

To me, right now, adoption is a four-letter word. It would be a Band-Aid. Right now adoption would not heal my frustration and pain and anger at being unable to get pregnant. To me having a child is about having a preg-

nancy. It's about telling my husband "I'm pregnant" and celebrating. It's about hearing the heartbeat for the first time and feeling the baby kicking inside of me. It's about procreating and going through the pains of labor and working with a doctor to successfully deliver. It's about my parents coming to the hospital to see the baby. It's about looking down at my kid and seeing my grandfather's expression in his face. To me it's about the whole process. Having an adopted newborn delivered to me would mean I missed a whole lot. I just couldn't do it. Our family is already complete with me and my husband. The only reason we would bring a child into our family would be to procreate, so that our blood lineage would be imprinted on this child. If IVF doesn't work for us, then we're going to have to grapple with the emotional decision of deciding not to parent.

Choosing childlessness is easier when both spouses agree on it. When it results from a disagreement over whether to choose nongenetic parenthood, it's much harder. Unfortunately, there isn't much room for compromise on an issue like this.

Infertility counselors can help you as you explore the option of childlessness. If you're a spiritual person and/or a member of an organized religion, a clergyperson or spiritual counselor may be able to help you put childlessness into perspective in your life. So will talking with other infertile couples who've chosen to remain childless, journal writing, and cognitive restructuring. Career counseling may help, too—sometimes patients who decide to remain childless make changes in their careers. Some are looking for ways to fill the emptiness left in their lives by childlessness; others decide that without children their careers are more important and thus must be more satisfying. Still others, recognizing the financial freedom of childlessness, choose careers or life changes that take advantage of that financial freedom.

Whatever you decide to do—continue infertility treatment in hopes of conceiving a genetic child, adopt a baby, become pregnant with the help of a donor, ask a surrogate to carry your child, or choose to remain childless—I hope that you will continue to use the many mind/body skills that you've learned in this book. These skills are wonderful for infertility, but they also come in handy when infertility is no longer an issue for you. Eliciting the relaxation response daily can benefit your mind and your body throughout your life. Minis help you cope with any stressful situation that flares up during the day, from being stuck in a traffic jam to the demands of a difficult boss. Cognitive restructuring, self-nurturance, journaling, paired listening, self-care, and all the other skills you've learned while battling infertility can help you live a healthier, less stressful, more peaceful life, now and forever.

APPENDIX ONE

Lifestyle Changes That Increase Fertility

In addition to the many mind/body strategies described in this book, lifestyle changes also can have a positive impact on your fertility. Exercising too much, weighing too little, even using some alternative health remedies can affect your ability to get pregnant. In this section I will share with you the advice I give all my infertility patients regarding lifestyle changes. I'll also tell you my opinions, based on the latest scientific evidence, regarding the alternative and complementary remedies that are sometimes used to treat infertility. Some of my stands are controversial—for example, your doctor may not agree with my recommendations on exercise—but I believe that the lifestyle changes I will describe here do contribute to the success my patients have in getting pregnant.

EXERCISE

There's quite a bit of controversy regarding the impact of exercise on fertility. Although there is consensus that extremely vigorous exercise leads to impaired fertility in women and men, there are continued debates about the impact of more moderate exercise on female fertility. Women who exercise very vigorously tend to stop ovulating and menstruating, and men who do so have decreases in testosterone production that can cause drops in libido and/or sperm production. Research has shown that infertile individuals who exercise strenuously often experience a return of their fertility when they reduce the intensity of their exercise.

The research on the impact of moderate exercise on female infertility is less clear. It has been hypothesized that some women are reproductively "exercise sensitive," and even moderate levels of exertion contribute to infertility. Progesterone production might be adversely affected by exercise, thus preventing embryo implantation.

I believe that exercising vigorously enough to bring your heart rate over 110 beats per minute can have a dampening effect on fertility. As a result, I advise my infertility patients to take a three-month exercise vacation. This is not an easy thing for many of my patients to hear—in fact, it is by far the most unpopular recommendation I make. But I have seen so many pregnancies occur in infertile women who simply stopped exercising that I feel compelled to continue making this recommendation.

Although many infertile women already feel that they've had to make too many life changes and sacrifices in their attempts to get pregnant, this one is temporary. If you decide to follow this advice, stop doing any exercise that brings your heart rate above 110 beats per minute for a three-month trial period. This means strolls, not power walks. It means no running or even jogging, no aerobics classes, no StairMaster. But gentle yoga is OK. If after three months you're still not pregnant, you can return to your normal exercise routine knowing that exercise probably isn't causing your infertility. If you do get pregnant, make sure to consult with your obstetrician before returning to your exercise routine.

BODY WEIGHT

To be able to reproduce, both women and men need a certain amount of body fat. Being underweight can cause ovulation to stop in women and can adversely affect sperm production in men.

In a study published in the mid-1980s, researchers looked at a group of women who were slightly below their ideal body weight and who were having difficulty getting pregnant. They were all advised to gain a small amount of weight. Surprisingly, half of the women refused to gain weight (which suggested that being thin was more important than having a baby, and which goes against all theories of evolution stating that procreation is the strongest instinct in the animal kingdom). Of the women who did choose to gain weight—an average of less than ten pounds—73 percent conceived spontaneously within three months. Other research has shown that women who are of normal weight but have to be extremely careful about maintaining it might have abnormal ovulatory patterns.

If you suspect or have been told that you are underweight, or if you find yourself having to be scrupulous about your eating patterns in order not to gain weight, you may well increase your chances of getting pregnant if you gain a few pounds. Remember that infertility is a temporary condition and that once you are finished with infertility, you can work on being as thin as you're comfortable being. But when trying to make a baby, being thin isn't necessarily best.

Ironically, being very overweight can also impair your ability to conceive and may even reduce the success of high-tech treatment. Several recent studies have shown that women with body-mass indices (BMI) greater than 27 do not respond as well to injectible fertility medications and have reduced pregnancy rates compared with slimmer women. However, losing even a moderate amount of weight—as little as 10 to 15 pounds—can return pregnancy rates to normal.

To determine whether you are under- or overweight, calculate your body-mass index in this way:

- Multiply your weight in pounds by 703.
- Divide that amount by your height in inches squared.

For example, if you are five foot five, weigh 140 pounds:

> 140 × 703 = 98420
> 98420 ÷ 4225 (65 squared)
> Your BMI is 23.3.

Women whose BMI is between 20 and 25 are the most fertile. If your BMI is under 17 or over 27, your chances of conceiving are significantly reduced. If your BMI is between 17 and 20, or between 25 and 27, you may be slightly less likely to conceive. Men with a BMI under 18 may also have reduced fertility.

WEIGHT-LOSS PLANS

If you are considering gaining or losing weight in an effort to increase your chances of conception, be extremely careful about what diet plan you follow. The best strategy is to consult a registered dietitian, who will design a healthy eating plan for you to follow. Many of the popular, trendy diets are completely untested and can have adverse effects on your health and perhaps your fertility.

There have been several books published recently claiming that if you follow a certain diet, you will significantly increase your chances of conceiving. Many of these books have been written by former infertility patients who followed some diet plan, conceived, and are sure that the diet facilitated the pregnancy. There has never been any scientifically valid research that documents such diets or that backs up the idea that adding or omitting specific food groups or products significantly increases your chance of pregnancy.

CAFFEINE

Several studies have linked caffeine consumption to reduced fertility as well as to an increased risk of miscarriage. In fact, a recent study suggests that high caffeine consumption before conception may contribute to miscarriages. Even a few cups of coffee a day can make a difference. In one study, women who consumed more than 500 milligrams of caffeine per day showed reduced fertility rates. Smokers who consume caffeine are even more likely to have trouble conceiving.

I advise my patients to eliminate caffeine completely. In order to do so without experiencing withdrawal symptoms, taper down consumption by half a cup a week. Male caffeine consumption doesn't appear to affect fertility.

Keep in mind that coffee isn't the only food that contains caffeine. Tea, soft drinks, and chocolate do, too. The following chart shows caffeine levels in a variety of foods:

Item (8-ounce serving)	Caffeine (milligrams)*
Starbucks regular drip coffee	170–240
Brewed coffee	65–120
Instant coffee	60–85
Espresso	30–50
Instant tea	24–31
Brewed tea	20–110
Cola	20–40
Iced tea	9–50
Cocoa	3–32
Chocolate milk	2–7
Decaffeinated coffee	2–4

*A range is given because the amount of caffeine varies based on how the beverage is made and what kind of coffee beans or tea leaves are used.

ALCOHOL

Until recently there was no evidence that alcohol consumption affected fertility. We knew that drinking during pregnancy put the baby at risk, but that was it. However, in 1999 two studies found that alcohol consumption can delay conception in *fertile* women. The researchers in this study found that the more alcohol fertile women drank per week, the longer it took them to get pregnant. And in a Danish study that looked at couples between the ages of twenty and thirty-five who were trying to conceive their first child, women who had more than ten drinks a week were half as likely to get pregnant as those who had fewer than five drinks a week.

No similar studies have been done with infertile women, but as far as I'm concerned, that's enough evidence to recommend abstinence for infertile women. A glass of wine now and then probably won't hurt, but if you have unexplained infertility, or if you're having infertility treatment, it's definitely worthwhile to quit drinking completely to see if it helps.

Moderate alcohol consumption doesn't seem to affect men. Those who drink a lot have trouble with impotence and loss of libido; moderate drinking appears to be OK. However, many of my patients have been quite pleased when their husbands, in a show of support, have altered their health habits as well.

SMOKING

Many studies support the claim that smoking can cut your chances of getting pregnant and carrying a pregnancy to term. For example, a recent study found that exposure to even a small amount of the chemicals in cigarette smoke may destroy a woman's eggs in the ovaries. What's more, smoking is suspected of contributing to early menopause, and women who smoke are less likely to have successful fertility treatments. One study found that in order to conceive, smokers require almost twice as many IVF cycles as nonsmokers. And some research indicates that male smoking might impair reproduction, although other studies fail to show that connection.

PRESCRIPTION AND OVER-THE-COUNTER MEDICATIONS

Drugs—even those sold over the counter—can sometimes interfere with sexual drive, male potency, and infertility medications. Be sure to tell your infertility doctor about any drugs you take, even those you buy over the counter.

HERBS

There is an assumption in this country that herbs are more beneficial than some medications since they are "natural." Natural or not, herbs can act like medicine, but unlike standard medications, their safety, effectiveness, and appropriate dosages are not adequately studied or regulated. A number of herbs may have adverse effects on your or your partner's reproductive system, and if you take them while pregnant, they may harm your unborn baby. The same goes for certain Chinese teas, the mixtures you make when a Chinese herbalist gives you a selection of herbs with which to make a tea.

In a recent study at Loma Linda University, four herbs were examined to determine their effect on fertility. The four herbs were St. John's wort, echinacea purpurea, ginkgo biloba, and saw palmetto. All of the herbs except saw palmetto made eggs harder to fertilize, reduced the viability of sperm, and altered the breast cancer gene, possibly increasing the risk to future offspring of developing breast cancer. Blue cohosh, an herbal dietary supplement often sold as a menstrual remedy, has been shown to cause severe birth defects in rats, including nerve damage and eye development problems.

Recent research also showed that St. John's wort decreases the efficacy of the birth-control pill, which is often used to prepare for an IVF cycle. Some scientists hypothesize that if St. John's wort can negatively impact the efficacy of a hormone-based medication such as the birth-control pill, it may also impair the efficacy of other medications, such as infertility drugs.

It is extremely important to let your infertility doctor know if you're taking any herbs at all. Your doctor will likely suggest that you stop all herbal remedies until you've completed your family.

ACUPUNCTURE

The ancient Asian art of acupuncture is commonly recommended for infertile individuals. Many acupuncture practitioners actually specialize in the "treatment" of infertility and have long waiting lists. Although there are numerous anecdotal reports of pregnancies following acupuncture treatment, there has been only one study documenting such an effect. In that study, 42 percent of the IVF patients who received acupuncture conceived, compared with 26 percent of the women who did not have acupuncture.

Acupuncture can be an effective treatment for psychological distress in some people. Since there is some evidence that depression may impair fertility (see Chapter 1), it's possible that in some individuals acupuncture treatment leads to reductions in depressive symptoms that may in turn facilitate increased pregnancy rates. It should be noted that this hypothesis has yet to be scientifically tested.

If you're interested in pursuing acupuncture treatment, I recommend that you approach it as a treatment for your psychological distress, not as an infertility treatment per se. Make sure to use a licensed, experienced acupuncturist, preferably one who has experience working with infertile individuals. And avoid any acupuncturist who guarantees pregnancy if you sign up for a certain number of sessions.

VITAMINS

Good nourishment can have an important influence on your fertility. The best way to make sure that you're eating a healthful, well-balanced diet, which is certainly advisable if you're trying to get pregnant, is to eat a varied menu rich in fruits, vegetables, and whole grains. Since that is sometimes a challenging task, I recommend that both men and women take a single multivitamin every day. *I caution you against taking megadoses of individual vitamins unless your physician or registered dietitian prescribes them.* And even then make sure your infertility specialist agrees with the prescription. Be especially careful of fat-soluble vitamins such as A and D. Vitamin A can actually cause severe birth defects when taken in high doses.

Folic acid is one of the B vitamins. It has been shown to help prevent birth defects in the brain and spinal cord. It is extremely important that you have enough of this vitamin in your system *before* you get pregnant. Prenatal vitamins tend to contain more folic acid than standard multivitamins do, so many infertility doctors advise their patients to take prenatal vitamins instead of regular multivitamins. The recommended dose of folic acid for women trying to get pregnant is 400 micrograms daily.

Calcium is another important part of your ideal daily diet. Women of childbearing age should get approximately 1,200 milligrams of calcium daily, which can be had by consuming the equivalent of three eight-ounce servings of milk (skim is your best choice). Since many women fall short on dietary calcium, supplements are a good idea. Be sure to avoid calcium derived from oyster shells or bone meal. They might contain lead, which can be harmful to an unborn child.

There is some evidence that zinc is important for the production of normal, healthy sperm. Make sure that your partner's daily multivitamin contains the recommended daily allowance (RDA) for zinc.

MASSAGE THERAPY, CRYSTALS, COLONIC IRRIGATION, AND THE LIKE

There are a huge number of alternative "remedies" offered to infertile women and men. Some, like massage therapy, are harmless, feel good to most people, and may be a very welcome self-nurturing treat during the crisis of infertility. Others, like crystals, are backed by no scientific evidence that they will help you in any way, but if you would like to use them, fine, there's apt to be no harm. However, there are other "treatments" offered to infertile individuals, such as colonic irrigation, that can in fact be harmful to your health. To my horror, one of my patients actually told me that there is a practitioner who guarantees pregnancy in women who come to her for colonic irrigation. Not only is this malpractice in every sense of the word, it is unethical to promise something to desperate patients with absolutely no evidence to support one's claims.

If you're interested in pursuing some form of alternative medicine, do so carefully. Be keenly suspicious of practitioners who make claims of pregnancy. And be sure to tell your infertility doctor about every treatment or remedy you pursue outside of his or her office.

APPENDIX TWO

Resources

The following organizations, Web sites, and books offer a wealth of information for infertile couples.

FOR INFORMATION ON INFERTILITY

Alice D. Domar, Ph.D., Director
The Mind/Body Center for Women's Health at Boston IVF
40 Second Avenue, Suite 300
Waltham, MA 02451
Phone: (781) 434-6500
Fax: (781) 434-6501
Web site: www.bostonivf.com
E-mail: conqueringinfertility@bostonivf.com

Boston IVF offers a variety of infertility mind/body services:

- **Ten-session programs.** The Mind/Body Program for Infertility is a comprehensive, complementary program designed to decrease symptoms, reduce isolation, and educate participants on the potential adverse impact of certain lifestyle behaviors on their reproductive health. Clinical outcomes for participants in the ten-session program show statistically significant decreases in physical symptoms, all measured psychological symptoms, and a 44 percent pregnancy rate.
- **Weekend retreats.** Using the same components of the ten-session program in Boston, this two-day retreat teaches individuals or couples a variety of relaxation techniques, stress-management strategies, and numerous behavioral and cognitive skills to lessen the stress of coping with infertility. Retreats are held at various times across the country.
- **Individual or couples psychotherapy.** I see a handful of patients on a weekly basis either because they don't want to be in a group, or they've gone through the group and want to discuss something privately, or they need some one-on-one attention in terms of a crisis brought on by infertility. I also see couples that are unable to make a treatment decision or are at odds about what to do next.
- **Personal consultations.** Women seeking immediate, personalized information can talk to me through our consultation service. During a sixty-minute telephone call or visit, I will address questions and concerns and offer advice on issues relating to infertility.
- **Professional training.** We offer training to licensed heathcare professionals on how to teach mind/body skills to infertility patients in a variety of formats and locations.
- **Relaxation materials.** For relaxation tapes and other materials, please look at bostonivf.com or mbmi.org.

Resolve, Inc.
1310 Broadway
Somerville, MA 02144
Phone: (617) 623-0744

E-mail: resolveinc@aol.com
Web site: www.resolve.org
Resolve is a fantastic resource for infertility patients. This nonprofit organization offers information, publications, and fact sheets on a wide range of topics, from the basic facts of infertility all the way to adoption. It also sponsors support groups, meetings, lectures, and symposia.

The American Society for Reproductive Medicine (ASRM)
1209 Montgomery Highway
Birmingham, AL 35216-2809
Phone: (205) 978-5000
ASRM's Web site (www.asrm.org) is a wonderful source. It offers a variety of frequently asked question (FAQ) worksheets, patient information booklets, patient fact sheets, referrals to professionals who do infertility counseling, information on state insurance laws, assisted-reproductive technology success rates, advice on selecting an IVF/GIFT program, and tips on finding a doctor.

The National Center for Health Statistics
6525 Belcrest Road
Hyattsville, MD 20782-2003
Phone: (301) 458-4636
Web site: www.cdc.gov/nchs/fastats/fertile.htm
NCHS provides statistics on infertility in the United States, statistics on assisted-reproductive technology success rates, and links to other government Web sites with infertility information. Other good government sites include the Centers for Disease Control (www.cdc.gov), the National Institutes of Health (www.nih.gov), and the U.S. Department of Health and Human Services' Healthfinder site (www.healthfinder.gov), which is an excellent guide to health issues of all kinds. In addition, the National Institute of Mental Health (www.nimh.nih.gov) provides excellent information on depression and other mental-health issues.

The Mind/Body Institute
UCLA Medical Plaza
13547 Ventura Blvd #642
Sherman Oaks, CA 91423

Phone: (213) 688-6119
Web site: www.mindbodyinfertility.com
I've trained health-care professionals all over the country in teaching mind/body techniques to infertility patients. The Los Angeles program, which is run by Katie Boland, is the most active program. For updates on other affiliated programs, check the Harvard Mind/Body Medical Institute Web site, www.mbmi.org.

BOOKS

Here is a handful of books that I like to recommend. This is by no means a complete list, however—there are dozens of helpful books about the medical aspects of infertility and other paths to parenthood. As you browse for books and other educational materials related to infertility, however, keep these two suggestions in mind: If you're interested in information about the medical aspects of infertility, choose a book written by a well-regarded health-care professional. And if a book guarantees pregnancy, I would be highly suspicious of it. As with so many other things in life, if it sounds too good to be true, it probably is—and whatever treatment the book is recommending may be dangerous.

Six Steps to Increased Fertility: An Integrated Medical and Mind/Body Program to Promote Conception, by Robert L. Barbieri, M.D., Kevin R. Loughlin, M.D., and me, Alice D. Domar, Ph.D. (Simon & Schuster/Fireside, 2001).

Choosing Assisted Reproduction: Social, Emotional, & Ethical Considerations, by Ellen Sarasohn Glazer (Perspectives Press, 1999).

Rewinding the Biological Clock: Motherhood Late in Life—Options, Issues, and Emotions, by Richard J. Paulson, M.D., and Judith Sachs (W.H. Freeman & Co., 1998).

Resolving Infertility, by Diane Aronson and the staff of Resolve (Harper Resource, 2001).

Girlfriend to Girlfriend: A Fertility Companion, by Kristen Magnacca (she was a patient of mine) (First Books, 2000).

Mommy, Did I Grow in Your Tummy? Where Some Babies Come From, by Elaine R. Gordon (E. M. Greenberg Press, 1992).

FOR INFORMATION ON ADOPTION

Adoptive Families of America
33 Highway 100 N.
Minneapolis, MN 55422
Phone: (800) 372-3300

Adoptive Families Magazine
42 West 38th Street, Suite 901
New York, NY 10018
Subscription Customer Service: (800) 372-3300
E-mail: letters@adoptivefam.com

American Academy of Adoption Attorneys
Box 33053
Washington, D.C. 20033-0053
Web site: www.adoptionattorneys.org

North American Council on Adoptable Children
1821 University Avenue #498 N.
St. Paul, MN 55104
Phone: (612) 644-0336

ODS Adoption Community of New England
(Formerly the Open Door Society of Massachusetts)
1750 Washington Street
Holliston, MA 01746
Phone: (508) 429-4260
Web site: www.odsma.org

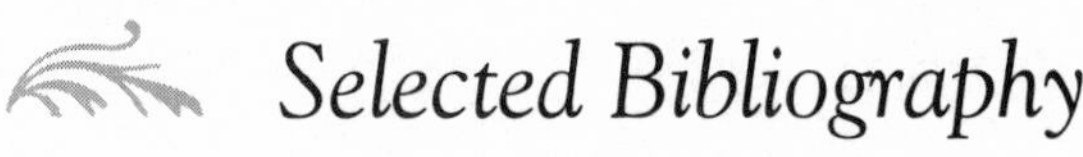

Selected Bibliography

For abstracts of most studies, you can refer to the National Library of Science's PubMed database at http://www.ncbi.nlm.nih.gov/PubMed/ and do a search on the title of the article or the authors' names.

BARAM, D., et al. Psychosocial adjustment following unsuccessful in vitro fertilization. J. Psychosom. Obstet. Gynecol. 8:181–90, 1988.

BEAUREPAIRE, J., et al. Psychosocial adjustment to infertility and its treatment: male and female responses at different stages of IVF/ET treatment. J Psychosom. Res. 38:229–30, 1994.

BELL, J. Psychological problems among patients attending an infertility clinic. J. Psychosom. Res. 25:1–3, 1981.

BOIVIN, J., AND TAKEFMAN, J.E. Stress level across stages of in vitro fertilization in subsequently pregnant and nonpregnant women. Fertil. Steril. 64:802–10, 1995.

CLARKE, R.N., et al. Relationship between psychological stress and semen quality among in vitro fertilization patients. Hum. Reprod. 14(3):753–58, 1999 Mar.

CONNOLLY, K.J., et al. An evaluation of counseling for couples undergoing treatment for in vitro fertilization. Hum. Reprod. 8:1332–8, 1993.

CREACH-LE MER, M.N., et al. Women's anxiety is a predictor of the implantation step of in vitro fertilization. Psychosom. Med. 61:92, 1999.

CURRIER, G.W., AND SIMPSON, G.M. Psychopharmacology: Antipsychotic medications and fertility. Psychiatr. Serv. 49:175–76, 1998.

DEMYTTENAERE, K., et al. Coping style and depression level influence outcome in in vitro fertilization. Fertil. Steril. 69:1026–33, 1998.

DOMAR, A.D., et al. The prevalence and predictability of depression in infertile women. Fertil. Steril. 58:1158–63, 1992b.

———, et al. The impact of group psychological interventions on pregnancy rates in infertile women. Fertil. Steril. 73(4):805–11, 2000 Apr.

———, et al. The impact of group psychological interventions on distress in infertile women. Health Psychol. 19(6):568–75, 2000 Nov.

———, FRIEDMAN, R., AND ZUTTERMEISTER, P.C. Distress and conception in infertile women: A complementary approach. 54:196–98, JAMWA 1999.

———, SEIBEL, M.M., AND BENSON, H. The mind/body program for infertility: A new behavioral treatment approach for women with infertility. Fertil. Steril. 53:246–49, 1990.

———, ZUTTERMEISTER, P.C., AND FRIEDMAN, R. The psychological impact of infertility: A comparison to patients with other medical conditions. J. Psychosom. Obstet. Gynecol. 14:45–52, 1993.

———, et al. Psychological improvement in infertile women after behavioral treatment: A replication. Fertil. Steril. 58:144–47, 1992a.

DOWNEY, J., AND MCKINNEY, M. The psychiatric status of women presenting for infertility evaluation. Am. J. Orthopsych. 62:196–205, 1992.

———, et al. Mood disorders, psychiatric symptoms, and distress in women presenting for infertility evaluation. Fertil. Steril. 52:425–32, 1989.

EUGSTER, A., AND VINGERHOETS, A.J.J.M. Psychological aspects of in vitro fertilization: A review. Soc. Sci. Med. 48:575–89, 1999.

FREEMAN, E.W., et al. Psychological evaluation and support in a program of in vitro fertilization and embryo transfer. Fertil. Steril. 43:48–53, 1985.

GALLETLY, C., et al. Improved pregnancy rates for obese, infertile women following a group treatment program. An open pilot study. Gen. Hosp. Psychiatry. 18(3):192–95, 1996 May.

GARNER, C., ARNOLD, E., AND GRAY, H. The psychological impact of in vitro fertilization. Fertil. Steril. Abstr. Suppl. 41:28, 1984.

HAMILTON, M. Development of a rating scale for primary depressive illness. Br. J. Soc. Clin. Psychol. 6:278–96, 1967.

HARLOW, C.R., et al. Stress and stress-related hormones during in vitro fertilization treatment. Hum. Reprod. 11:274–79, 1996.

HUNT, J., AND MONACH, J.H. Beyond the bereavement model: the significance of depression for infertility counseling. Hum. Reprod. 2:188–94, 1997.

JUDD, L., et al. Subsyndromal symptomatic depression: A new mood disorder? J. Clin. Psychiatry. 55:18–28, 1994.

KEE, B.S., JUNG, B.J., AND LEE, S.H. A study on psychological strain in IVF patients. J. Assist. Reprod. Genet. 17(8):445–48, 2000 Sep.

KLOCK, S.C., AND GREENFELD, D.A. Psychological status of in vitro fertilization patients during pregnancy: a longitudinal study. Fertil. Steril. 73(6):1159–64, 2000 Jun.

KLONOFF-COHEN, H., et al. A prospective study of stress among women undergoing in vitro fertilization or gamete intrafallopian transfer. Fertil. Steril. 76:675–87, 2001.

LAFFONT, I., AND EDELMANN, R.J. Psychological aspects of in vitro fertilization: a gender comparison. J. Psychosom. Obstet. Gynaecol. 15:85–92, 1994.

LAPANE, K.L., et al. Is a history of depressive symptoms associated with an increased risk of infertility in women? Psychosom. Med. 57(6): 509–13, 1995 Nov-Dec.

LEIBLUM, S.R., KEMMANN, E., AND LANE, M.K. The psychological concomitants of in vitro fertilization. J. Psychosom. Obstet. Gynecol. 6:165–78, 1987.

LUKSE, M.P., AND VANCE, N.A. Grief, depression, and coping in women undergoing infertility treatment. J. Obstet. Gynecol. 9:245–51, 1999.

MATIKAINEN, T., et al. Aromatic hydrocarbon receptor-driven Bax gene expression is required for premature ovarian failure caused by biohazardous environmental chemicals. Nat. Genet. 28(4):355–60, 2001 Aug.

MCGUIRE, L., KIECOLT-GLASER, J.K., AND GLASER, R. Depressive symptoms and lymphocyte proliferation in older adults. J. Abnor. Psychol., vol. 111, no. 1, 192–97, 2002.

MERARI, D., et al. Psychological and hormonal changes in the course of in vitro fertilization. J. Assist. Reprod. Genet. 9:161–69, 1992.

NESS, R.B., et al. Infertility, fertility drugs, and ovarian cancer: A pooled analysis of case-controlled studies. Am. J. Epidem. vol. 155 no. 3, 217–24, 2002.

NEWTON, C.R., HEARN, M.T., AND YUZPE, A.A. Psychological assessment and follow-up after in vitro fertilization. Fertil. Steril. 54:879–86, 1990.

ODDENS, B., DEN TONKELAAR, I., AND NIEUWENHUYSE. Psychosocial experiences in women facing fertility problems: a comparative study. Human. Reprod. 14:255–61, 1999.

READING, A.E., CHANG, L.C., AND KERIN, J.F. Psychological state and coping styles across an IVF treatment cycle. J. Repro. Med. 7:95–103, 1989.

SERRATTI, A., et al. Delineating psychopathologic clusters within dysthymia: a study of 512 outpatients without major depression. J. Affective Disorders 56:17–25, 1999.

SHAW, P., JOHNSTON, M., AND SHAW, R. Counselling needs, emotional and relationship problems in couples awaiting IVF. J. Psychosom. Obstet. Gynaecol. 9:171–80, 1988.

SMEENK, J., et al. The effect of anxiety and depression on the outcome of in vitro fertilization. Hum. Reprod. 16:1420–23, 2001.

THIERING, P., et al. Mood state as a predictor of treatment outcome after in vitro fertilization/embryo transfer technology (IVF/ET). J. Psychosom. Res. 17:481–91, 1993.

VISSER, A., et al. Psychosocial aspects of in vitro fertilization. J. Psychosom. Res. 15:35–43, 1994.

WILLIAMS, K.A., et al. Evaluation of a Wellness-Based Mindfulness Stress Reduction intervention: a controlled trial. Am. J. Health Promot. 15(6):422–32, 2001 Jul-Aug.

Index